Sediment Transport Research - Further Recent Advances

Edited by Andrew J. Manning

Published in London, United Kingdom

Sediment Transport Research - Further Recent Advances
http://dx.doi.org/10.5772/intechopen.1001488
Edited by Andrew J. Manning

Contributors
Andrew J. Manning, Besik Saralidze, Clara Bianchi Bahia, Daniela Glueck, Daniel E. Canfield Jr, Daniele Toyama, David Sakhvadze, Denise Balestrero Menezes, Eric Wolanski, Gigo Jandieri, Giorgi Sakhvadze, Hendrik Schubert, Jarrell Smith, Judy Q. Yang, Michael Elliott, Mina Kiani, Olga Tammeorg, Priit Tammeorg, Sivaramakrishnan Balachandar, Soukaina Benaich, Susumu Onaka, Takaaki Uda, Timothy J. Canfield, Tomohiro Mori, Xiao Yu

First published in London, United Kingdom, 2024 by IntechOpen
IntechOpen is the global imprint of INTECHOPEN LIMITED, registered in England and Wales, registration number: 11086078, 167-169 Great Portland Street, London, W1W 5PF, United Kingdom

British Library Cataloguing-in-Publication Data
A catalogue record for this book is available from the British Library

Additional hard and PDF copies can be obtained from orders@intechopen.com

Sediment Transport Research - Further Recent Advances
Edited by Andrew J. Manning
p. cm.
Print ISBN 978-0-85466-554-9
Online ISBN 978-0-85466-553-2
eBook (PDF) ISBN 978-0-85466-555-6

For EU product safety concerns:
IN TECH d.o.o., Prolaz Marije Krucifikse Kozulić 3, 51000 Rijeka, Croatia,
info@intechopen.com or visit our website at intechopen.com.

Meet the editor

Professor Andrew J. Manning is a technical director (Rank Grade 10) and professor in the Coasts & Oceans Group at HR Wallingford (UK) and has over 27 years of scientific research experience (in both industry and academia) examining natural turbulent flow dynamics and fine-grained sediment transport processes and assessing how the these interact (including both field studies and controlled laboratory flume simulations). Andrew is also an associate professor in coastal and shelf physical oceanography at the University of Plymouth (UK). Internationally, Andrew has been appointed visiting, adjunct, or guest professor at a number of universities (Hull, UK; Stanford, USA; Delaware, USA; Florida, USA; TU Delft, Netherlands; Loughborough, UK) and is a highly published and world-renowned scientist in the field of depositional sedimentary flocculation processes. He is a fellow of the Royal Geographical Society, was awarded a University of Plymouth Vice Chancellor's Research Fellowship (2007), and was presented the Exemplary Act Award by the U.S. Department of the Interior and U.S. Geological Survey (2015). Andrew has contributed to more than 110 peer-reviewed publications in marine science, which have been published in international scientific journals and books, plus over 160 articles in refereed international conference proceedings, and currently has an H-index of 31. He supervises graduates, postgraduates, and doctoral students focusing on a range of research topics in marine science. Andrew has led numerous research projects investigating sediment transport and dynamics in aquatic environments around the world with locations including estuaries, tidal lagoons, river deltas, salt marshes, intertidal areas, coastal waters, and shelf seas. Andrew actively promotes disability equality through his work and fundraising for the charity Scope, primarily by running in international marathons.

Contents

Preface

Sediment Transport Research – Further Recent Advances is a book that covers a wide range of topics. The effective management of many aquatic environments requires a detailed understanding of sediment transport and dynamics. This has both environmental and economic implications, especially where there is any anthropogenic involvement. Numerical models are often the tool used for predicting the transport and fate of sediment movement in these situations, as they can estimate the various spatial and temporal fluxes. However, the physical sedimentary processes can vary quite considerably depending upon whether the local sediments are fully cohesive, noncohesive, or a mixture of both types. For this reason, for more than half a century, scientists, engineers, hydrologists, and mathematicians have all been continuing to conduct research into the many aspects which influence sediment transport. These issues range from processes such as scour, erosion, and deposition to how sediment process observations can be applied in sediment transport modeling frameworks. This book draws on international scientific research to examine the following sediment transport–related issues: sand nourishment, mud dynamics, computational fluid dynamics (CFD), beach morphology, anthropogenic activities, soil erosion, dredging, flocculation processes, and water quality aspects. The research was carried out by researchers who specialize in the transport of sediments and related issues.

It is a great pleasure to write the preface to this book published by IntechOpen. It comprises eight chapters written by a truly international group of research scientists who specialize in areas such as sediment properties and characteristics, geomorphology, water quality, rheology, sedimentary related legislation, particle dynamics, and numerical modeling. The majority of the chapters are concerned with sediment transport–related issues in estuarial, coastal, or freshwater environments. For example, they present an assessment of suspended sediment dynamics during sand nourishments and storm events; and the impact of salt on cohesive sediment transport. Other contributions in this book include physics–biology links in mud dynamics, soil erosion and losses, turbulent flow assessments, relating beach morphology to waves and ground subsidence, power technology, and linking anthropogenic activities to sediment accumulation rates. Authors are responsible for their views and subsequent concluding statements.

In summary, this book provides an excellent source of information on recent research on sediment transport, particularly from an interdisciplinary perspective.

I would like to thank all of the authors for their contributions, and I highly recommend this textbook to both scientists and engineers who deal with the related issues.

Andrew J. Manning
Professor,
University of Florida,
Gainesville, Florida, US

HR Wallingford,
Wallingford, UK

Chapter 1

Impact of Salt on Cohesive Sediment Transport

Soukaina Benaich and Judy Q. Yang

Abstract

The erosion and transport of cohesive sediment are more difficult to study than non-cohesive sediment, largely because these processes vary with the salt in the water. Clay minerals are the major components that contribute to the cohesiveness of cohesive sediment because they have significantly larger surface charges and surface area-to-volume ratio than non-cohesive sediment. The electrochemically active clay surfaces can adsorb ions on their surfaces, form an electrical double layer, and cause clay particles to aggregate or form a gel. In this chapter, we first discuss the properties of clay minerals, including the structure of clay primary particles, their surface charge and area, and their interaction with ions in water. The surface charges and surface areas of clay are several orders of magnitude larger than non-cohesive sand, thus predisposing it to interactions with salt in aqueous environments. Second, we summarize studies that reveal the role of salts, specifically salinity and sodium absorption ratio (SAR), on sediment aggregation, stability, and settling speed. An increase in salinity from 0.15 to 1.5 ppt has been shown to increase the erosion threshold of smectite clay by more than 10 times. These findings underscore the crucial role of salt in shaping cohesive sediment transport.

Keywords: cohesive sediment, clay, mud, transport, salinity, ions, salt, sodium adsorption ratio

1. Introduction

A fundamental understanding of sediment transport and the effects of salts is essential for managing water resources, protecting habitats, and addressing concerns. Clay minerals, widely present in cohesive sediment, can interact with salts and sediment movement. In the following section, we summarize the characteristics of clay minerals including their classification, particle size distribution, surface area, and surface charges. These characteristics dictate how clay particles aggregate and interact with ions in water. We then explore the formation and significance of the electrical double layer (EDL) surrounding clay particles. We discuss the role of EDL in particle aggregation, dispersion, and transport. In the third section, we delve into studies that investigate how salinity influences sediment transport processes. We analyze how variations in salt concentrations affect factors like coagulation processes, stability levels of sediments, particle size distributions, and settling rates of clay aggregates. In Section 4, we examine the effects of sodium adsorption ratio (SAR) and sodicity on erodibility to underscore their significance in soil erosion processes across soil types. In Section 5, we delve into research elucidating the correlation between salt content and the gelatinous status of smectite clay, providing insight into the influence of salt quantity on clay erosion thresholds. Finally, we present viewpoints on how salt interacts with clay minerals and impacts sediment transport in aquatic environments.

2. Clay minerals

2.1 The properties and classifications of clay minerals

Cohesive sediment is ubiquitous in aquatic ecosystems, and clay is a major component of cohesive sediment [1]. In many fields such as geomorphology, clay is defined as fine sediment particles less than 2 microns [2]. Many of these micron-size clay particles, obtained by sieving the soils and sediment collected from the field, are made up of thousands to millions of nanometer-size clay primary particles, which are aluminosilicates that have layered structures. Clay minerals are a major contributor to the cohesiveness of sediment aggregates due to their large surface-area-to-volume ratio and large surface charges. Many clay minerals, such as kaolinite and smectite, consist of SiO_4 tetrahedral and AlO_4 octahedral structures, and they can be classified into two types based on how the tetrahedral and octahedral sheets are arranged, as shown in **Figure 1**. The first type is called the 1:1-layer type or silicate clay, which consists of one tetrahedral sheet bonded to one octahedral sheet through hydrogen bonds. This type of clay has a typical particle size of less than 10 μm, and a surface charge ranging from 0.5 to 9×10^4 m^2 kg^{-1}. An example of this type is kaolinite. The second type is known as the 2:1 layer type or smectite clay and it includes one octahedral sheet positioned between two tetrahedral sheets. Both illite and smectite clay have a 2:1 layer structure. The mean thickness of the smectite primary particle is on the order of 1 nm [3]. The structure of montmorillonite, a common smectite clay, is shown in **Figure 1**. The total surface area of smectite is about 700 m^2 g^{-1}.

In general, clay minerals can be further classified into five groups based on cation substitutions occurring within their layers. These groups include kaolinite (chemical composition: $Al_2Si_2O_5(OH)_4$), illite ($M_x[Si_{6.8}Al_{1.2}]$ $Al_3Fe_{0.25}Mg_{0.75}O_{20}(OH)_4$), vermiculite $M_x[Si_7Al]$ $Al_3Fe_{0.5}Mg_{0.5}O_{20}(OH)_4$, smectite ($M_x[Si_8]Al_{3.2}Fe_{0.2}Mg_{0.6}O_{20}(OH)_4$),

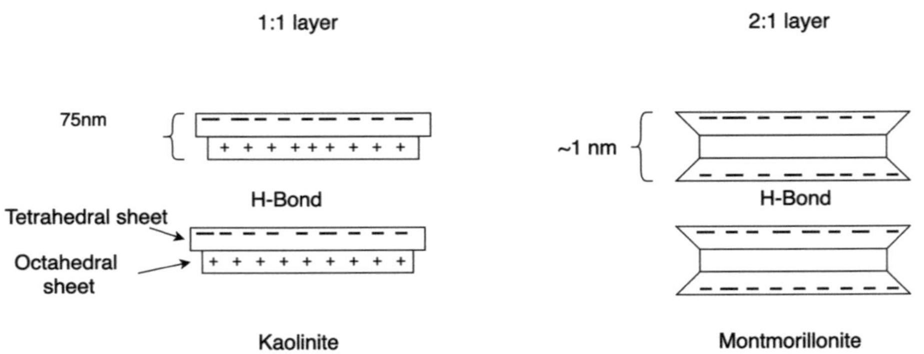

Figure 1.
Schematic diagrams of kaolinite and montmorillonite, which represent 1:1 and 2:1 clay types. Their sizes are indicated in the figure.

and chlorite ($(Al(OH)_{2.55})_4$ $[Si_{6.8}Al_{1.2}]$ $Al_{3.4}Mg_{0.6}O_{20}(OH)$). Each group has different characteristics and plays an important role in geological and environmental processes [4]. In this article, we focus on discussing kaolinite and smectite because they are the two most common clay and represent the 1:1 and 2:1 clay types.

2.2 The size distribution of clay primary particles

Most of the clay minerals are nano- to micro-size particles. For example, the lengths of kaolinite particles have three peaks, located at 0.075, 0.3, and 1.2 μm [5]. The smallest peak, which makes up 10–15% of the particles may indicate the thickness of kaolinite particles. The other two peaks correspond to the dimensions of the kaolinite's basal plane resulting in aspect ratios of 16 and 4, respectively [5]. The other two dimensions of smectite clay have been shown to range from 100 to 200 nm [4]. Ca^{2+} montmorillonite, a typical smectite clay, demonstrates two particle sizes reaching their peak at 0.6 and 1.5 μm while Na^+ montmorillonite shows a single prominent particle size with the highest occurrence, at 1.5 μm. These distributions provide information about the size features of montmorillonite [5].

2.3 Surface area and charges

Clay aggregates have extensive internal surface area because they consist of layers of clay primary particles. These layers of some clays such as kaolinite and montmorillonite are structured like a book, leading to significantly more internal surface area. For example, the surface area of kaolinite ranges from 169 to 170 m^2/g [6], which is about three orders of magnitude larger than the surface area of non-cohesive sand of a similar weight.

The surfaces of many clay minerals are mostly electronegatively charged. However, this might change depending on the chemical composition and pH of the surrounding solution. Specifically, the clay surface attracts thousands of cations like Na^+, H^+, and Ca^{2+}, which induces an ionic double-layer [7]. The electric charge, on pristine kaolinite clay nanoparticles measured through electrostatic force microscopy (EFM) is around 25 nC/cm^2 [8]. Estimation of the surface charge of clay is difficult because the clay can exchange cations with the solution and change its surface charge. For example, the surface charges of kaolinite are pH-dependent—its surface charge ranges from 2.9×10^{-14} to 6.4×10^{-15} nC/cm^2 for pH between 4 and 6 and becomes negative (-4.5×10^{-14} nC/cm^2) at pH = 9 [9]. Due to these large surface charges, clay can adsorb abundant cations on their surfaces and exchange abundant cations with the surrounding solution. The interactions between clay and cations are linked with the cation exchange capacity (CEC) of soils.

The interactions between clay and cations also depend on the type of cations because the interactions between clay minerals and multivalent cations are different from their interactions with monovalent cations. Multivalent cations can bridge two negatively surfaces charged surfaces. For example, a calcium cation with two positive charges can bind to a negatively charged clay particle on one side and capture another negatively charged clay particle on the other side, causing these clay particles to aggregate [10]. Bridging clay particles is a key mechanism behind the flocculation phenomenon [11]. Clay's retention and aggregation have been consistently shown to increase in the presence of divalent or multivalent ions [11]. In contrast, monovalent ions enhance the dispersion and spread of clay. Due to the complex interactions between clay and ions, the aggregation and transport of cohesive clay are expected to be largely impacted by the amount and type of ions in water [11].

2.4 Electric double-layer

When the particle is negatively charged, it attracts positive cations from the solution. This layer of cations is tightly bound to the surface and such layer is called the Stern layer, as shown in **Figure 2**. The cations in the Stern layer further attract anions in the solution that are less strictly adhered to the clay. This layer of anions is called the diffuse layer. The surface charge, Stern layer, and diffuse layer constitute the electric double-layer (EDL) [12]. EDL plays a critical role in the electrostatic stabilization of colloids and is fundamental for colloid stability and transport [13].

The interactions between EDLs play a critical role in particle aggregation and dispersion, and as a result, the factors that impact EDL will impact particle aggregation, dispersion, and transport. The surface potential of clay and the disjoining pressure among clay are two commonly used parameters that characterize the electric potential on the surface of a particle and the electrical interactions among two clay particles. Both parameters are related to EDL and are impacted by many factors including clay surface charge, the distance between clay particles, the polarization of water, and the ions in the solution. Surface potential is influenced by the surface charge and the separation between surfaces when EDLs overlap. In addition, according to the Poisson-Boltzmann theory, the polarization of water molecules and counterions influences how ions/water interact with the soil particle surface and can cause a reduction in surface potential. The midpoint potential between colloidal particles decreases with increased ion polarization [14].

Furthermore, the electrostatic part of disjoining pressure is linked to the interactions of the layer (EDL). Water molecule and counterion polarization impact the relationships between ions/water and soil particle surfaces, which impact the disjoining pressure. Water molecule polarization enhances pressure among soil particles whereas counterion polarization diminishes it. The interactions within the EDL affected by water and counterion polarization have a significant impact on how soil colloidal particles come together or separate and their transport [14]. Furthermore, the EDL is impacted by the types of ions in water. Divalent ions have been found to decrease the double-layer thickness, which could lower the colloid surface potential

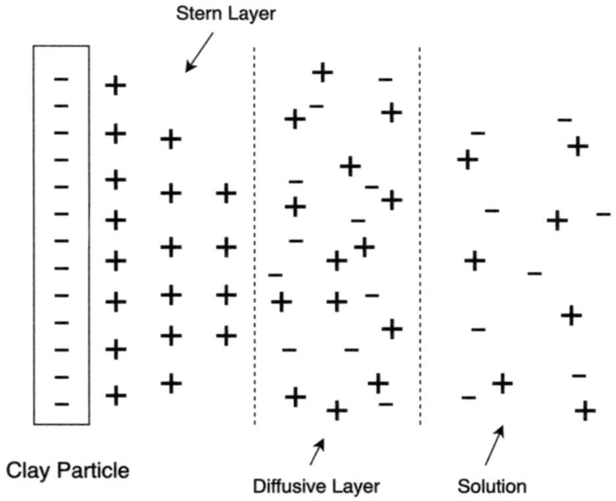

Figure 2.
Schematic diagram of the electrical double layer surrounding a negatively charged clay particle. The size is not to scale.

and impact colloidal transport. Knowledge of the impact of ions on EDL and clay disjoining pressure could be used to improve understanding of the role of salt on clay aggregation, dispersion, and transport. However, current knowledge of the EDL of clay and the nano- to microscale clay interactions remains incomplete, making it difficult to understand how salt impacts clay aggregation, dispersion, and macroscale transport. In the next sections, we present studies demonstrating how salt impacts clay and cohesive sediment transport at the macroscale.

3. Salinity and sediment transport

Salts have been shown to impact the transport of cohesive sediment at the macroscale (centimeter to meter scale). Such impact can be attributed to the large surface charge and area of clay and its ability to adsorb cations and form an EDL, as described above. In this section, we summarize papers that discuss how salinity impacts cohesive sediment transport. Salinity (g/L) is a key parameter that represents the total mass of dissolved salt in a liter of water. Salinity has been shown to impact the coagulation and stability of clay Edzwarld et al. [15], the growth rate and shape of the aggregates, and settling velocity [16], as discussed below.

First, Edzwarld et al. [15] studied the connection between ionic strength and the coagulation of suspended clay minerals through laboratory experiments and field investigation. For laboratory experiments, they mixed three types of clay minerals (kaolinite, illite, and montmorillonite) with buffer solution at ionic strength ranging from 0.03 to 0.3 M. Afterward, they placed the solution in a reactor with a stirring shaft that generates controlled shear and measured the number of particles under shear using microscopic counting as a function of flocculation time. Their study shows that the number of particles decreases with increasing flocculation time, which is on the order of 10 minutes. They further show that the number of particles decreases faster with increasing ionic strength, suggesting that an increase in ionic strength increases the flocculation of these three types of clay. From their number of particles versus the time curve, they determined the stability factor based on a mathematical model developed by Smoluchowski [17]. Specifically, Smoluchowski [17] described the aggregation of particles due to the movement of the fluid.

$$\frac{dn}{dt} = \frac{-4\Phi\alpha Gn}{\pi} \tag{1}$$

Here, n is the concentration of particles at time t, α is the stability factor, ϕ is the volume of colloidal particles per unit volume of suspension, and G is the root mean square velocity gradient. This equation demonstrates that the aggregation of clay particles is controlled by the shear as well as the stability of the factor, α. Edzwarld et al. [15] estimated α based on the measured n versus t curve for these three types of clay mixed with a solution at various salinity, and their laboratory experiments show that α increased by 2–4 times for all three clays when the ionic strength increased from 0.05 to 0.3 M. They further show that this trend held across solution compositions whether it was buffered NaCl solutions or synthetic estuarine solutions. This finding emphasizes an influential connection between ion concentration and the rate of coagulation.

In addition to laboratory experiments, Edzwarld et al. [15] collected sediment samples from the Pamlico River Estuary, which has a salinity ranging from 0.5 ppt in

Washington to 16 ppt at the mouth of the river. Afterward, they determined the type of clay minerals in the sediment samples using X-ray diffraction analysis. They also measured the number of particles in a reactor with a stirring shaft, from which they estimated the stability value α. Their results show that the stability factor of the field clay samples increased by 2–4 when salinity increased from 0.04 to 0.3 M. This result further confirms that salinity plays a critical role in clay coagulation.

In addition to coagulation and stability factors, salinity has been found to impact the size and shape of clay colloids as well as how fast they deposit. Abolfazli and Strom [16] conducted laboratory experiments to investigate how changes in salinity impact the aggregation of three types of clay: kaolinite, natural mud, and natural mud without organic matter (treated with sodium hypochlorite). The experiments were conducted in a 13-L mixing chamber, and turbulence was generated by a stirrer motor connected to a paddle, which allowed for adjustment of the mixing rate and turbulent shear rate (G) within the chamber. The mixing chamber had a camera system integrated into its setup to measure the size of particles or flocs. This system consisted of an LED source placed inside the tank and an external camera that captured images of particles passing through the opening created by the light source and tank wall. The camera system is capable of capturing flocs ranging from 5 to 1500 μm in size. The sizes of these particles or flocs were determined by extracting their projected two-dimensional area (A_f) using the formula $df = \sqrt{\frac{4A_f}{\pi}}$ where d_f represents the diameter or size. To generate size distributions based on both volume and size statistics like d_{50} (which represents the floc size by volume) the identified particles were categorized into log-sized bins. Abolfazli and Strom [16] measured d_{50} at various ionic strength and shear rates. In addition, they calculated the growth rate of the flocs as the changes in d_{50} during a short period (D_t) after a stepwise decrease in turbulent shear rate. Based on the size and growth rate measurements, they found that for mud with and without organic matter, an increase of salinity by 1–3 ppt can decrease the number of non-flocculated particles and increase floc size. Their study indicates a strong interaction between organic matter and salt, which needs further investigation.

In addition to the flocculation stability, salinity has been shown to impact the size and speed at which sediment particles settle, key parameters that control suspended sediment transport [18]. In this study, cohesive-suspended sediment was sampled from the Severn Estuary, high shear was first applied to break down these sediments, and then low shear was applied to suspend the particles for two hours. Afterward, the distribution of particle sizes and the settling velocity of the particles were measured using particle image velocimetry and image processing techniques. The results showed that sediment particles tend to clump when there is a higher concentration of sediment, but this change in the size of the flocs is not affected by changes in salinity levels. The speed at which particles settled ranged from 0.2 to 1.2 mm/s and was influenced by both the concentration of suspended sediment and salinity levels. The study discovered that in saltwater when sediment concentration is higher, the proportion of larger flocs increases. However, at lower salinity levels (<2.5 ppt), this proportion decreases by 7%. When salinity is low (2.5 ppt), the settling velocity increases with increasing suspended sediment concentration. However, at higher salinity levels (20 ppt), the floc settling velocity decreases with increasing suspended sediment concentration. The research suggests that salinity plays a key role in the setting of suspended sediment.

Krahl et al. [18] conducted a study to explore the role of salinity and extracellular polymeric substances (EPS) on the settling behavior of smectite clay, Wyoming bentonite. The main goal was to determine the coagulation concentration (CCC) needed for bentonite to start forming flocs in saline water. The researchers used the following three setups: differential settling experiments to measure CCC, settling time standard hydrograph analysis to study grain size distribution, and jar test experiments to compare floc size and settling behavior of bentonite with and without xanthan gum (a substitute for EPS). Sodium chloride was used to adjust the salinity and simulate neutral pH conditions in riverine waters. The findings indicated that even a small amount of salinity causes aggregation. For aggregates with a size of 184.8 μm, a settling speed of 2.0 mm/s was measured. When EPS was added to bentonite in saltwater, the aggregate sizes increased and settled faster than those without it. They further suggest that salinity is the trigger of flocculation, while EPS can further increase aggregate size. These findings further highlight the importance of salinity in influencing how particles clump together and the settling of these particles, as well as the complicated interactions among organic matter and salts. Investigations of the interactions between organic matter, salt, and clay minerals are needed yet currently lacking.

4. Sodium-adsorption ration and sodicity

In addition to salinity, the sodium absorption ratio (SAR), which is the amount of sodium Na relative to calcium (Ca) and magnesium (Mg) in water extract from saturated soil, plays a significant role in sediment erodibility, yet its role has been under-researched [1]. Soils with high SAR cause the clay to have higher water absorption and high porosity, which could lead to high erodibility. Parwada et al. [19] investigated how various physical and chemical factors affect the erosion rate of soils, including the SAR. They prepared clay samples with varying compositions of pore fluids. These samples were compressed into tubes and allowed to expand before being tested in a rotating cylinder apparatus. Then they placed the samples in a rotating cylindrical device and measured the erosion rate under different shear stresses. The results suggest that the types of cations, for example, monovalent versus multivalent, have a noticeable impact on sediment erodibility, yet such impact remains to be quantified.

Sodicity, defined as high sodium ion concentration compared to other cations in soil, has been used to characterize the impact of sodium on soil erosion. When soil structure starts to get impacted because of high sodium concentration, we call it sodic soil. Parwada et al. [19] collected soil samples from the Ntabelanga area in Eastern Cape province, South Africa, and analyzed the SAR, clay dispersion ratio, and other analyses to reveal a correlation ratio with land loss. The study found that clay presence contributes to soil stability, soils with more sand or silt tend to be more erodible. Soil sodicity, on the other hand, was measured by exchangeable sodium percentage (ESP) and SAR parameters, indicating that it can have an impact on soil erosion. However, in the Ntabelanga area, the sodicity level was low, and it was not the main cause of soil erosion, compared with other factors such as the soil organic carbon content. Such observations further highlight the importance of organic-salt interactions in clay erodibility.

Knowledge about the role of salts on erosion is particularly important for countries like Australia, where the salinization of water and land is significant and has been shown to cause land loss. It has been estimated that about 5.7 million hectares of soil in Australia are at stake and up to 20,000 km of streams are estimated to be affected 17 years from now [20]. Previous studies have linked high sodium levels with increased erosion and sediment transport, yet little evidence exists to support the relationship between sodicity and soil erodibility [20]. Ghadiri et al. [20] conducted a study that compared two soils with different characteristics, one with and another one without sodium. The research investigates the effects of soil salinity and sodicity, on soil erodibility, sediment transportation, and water quality downstream. Through measurements of erosion in simulated rainfall conditions on two soils in Queensland, it was observed that sediment loss increased when the soil was treated with sodium. These observations suggest that higher sodicity can lead to a weakening of soil aggregates and that the type of ions plays a critical role in sediment and soil erosion.

Like sodicity, Singer et al. [21] investigated the effects of Exchangeable Sodium Percentage (ESP) on soil erodibility in the Colusa Basin drain. They conducted three experiments using a rainfall simulator in their laboratory. The experiments involved eroding samples of soil types, such as Colusa silt loam, Pescadero, Sacramento A horizons, Contra Costa silty clay loam, and Hillgate loam. To measure ESP, Singer et al. experimented using soil samples, prepared soil batches, and applied leaching solutions with varying ratios of NaCl and $CaCl_2$. These solutions were then used to wash moistened soil samples weighing 1100 grams each. After leaching the amount of sodium was determined, in the soil. For Hillgate soil, the values were 16, 42, 69, and 80%. For Contra soil, the values were 19, 42, 75, and 81%. To remove salts from the samples a solution containing 95% ETOH was used, which was confirmed by conducting an $AgNO_3$ test on the leachates. The actual ESP was determined by displacing sodium, with NH_4 acetate. To ensure uniformity in the effect of leaching solutions divalent cations were first removed using Na acetate before leaching the soil using solutions that contain Ca and Na ions. Soil loss, soil detachment, and runoff volume were measured to assess how ESP influences soil erodibility. They first observed that lower ESP values (4) in Colusa silt loam resulted in a reduction of 20–24% in soil loss compared to high ESP values (54 and 73) [21]. These results highlight the impact of exchangeable ions on soil erosion.

Carlin et al. [22] conducted a field investigation in the Gulf of Mexico to examine the salt wedge's movement and the sediment's transport from the Brazos River to the Gulf of Mexico. A salt wedge is a layer of salt or seawater below fresh water. They measured salinity and turbidity and collected bottom topography and seismic data to investigate how a salt wedge interacts with suspended sediment in the Brazos River. To profile the water column, CTD (conductivity, temperature, and depth) and OBS (optical backscatter sensors) probes were deployed to measure the salinity and suspended sediment concentration during research expeditions in 2007, 2010, and 2012. The concentration profiles measured by OBS both longitudinally and vertically were analyzed to create a model representing sediment concentration across the river section. A mass balance analysis was carried out to estimate changes in transport from loads to the Gulf of Mexico. Their results show that the salt wedge modulates the transport of suspended sediment. Specifically, they show that when river discharge is low, it results in a salt wedge that captures sediment and creates a decimeter-thick mud layer. These field investigations further emphasize the important role of salt in modulating the transport of cohesive sediment, especially in estuaries where freshwater and saltwater meet.

5. Salinity and clay gelatinous status

San Juan et al. [23] investigate how salinity impacts the erosion threshold and microstructures of smectite clay. The study examined the erosion threshold of clay using an oscillating-grid tank having an oscillating grid moving up and down to generate a range of shear stress and turbulent kinetic energy. They tracked the movement of a synthetic smectite clay, laponite, using a laser sheet and a digital camera. In addition,

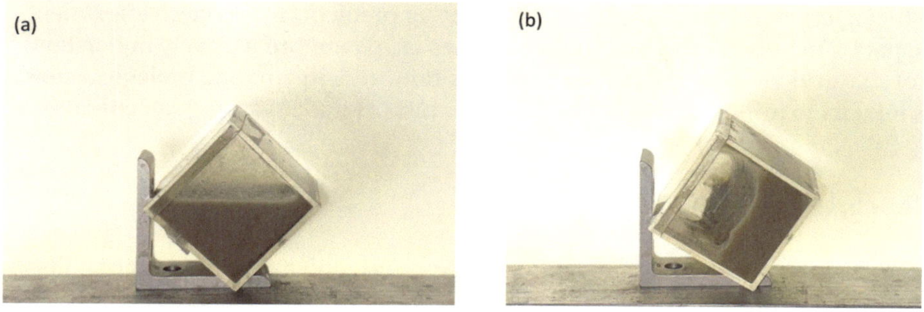

Figure 3.
Pictures of bentonite samples prepared at salinity 0 ppt (a) and 35 ppt (b). The sample was tilted by 30° to illustrate the gel or liquid states.

they directly measured the shear stress and turbulent kinetic energy using particle image velocimetry. They gradually increased the shear and identified the critical shear stress to erode the clay. The experiments were repeated at seven water salinities ranging from 0.1 to 10 ppt, as well as two different consolidation states. In addition to erosion measurements, they measured the yield stress of the clay samples using a rheometer.

Furthermore, they imaged the microstructure of the clay under various salinities using a confocal laser scanning microscope. The study discovered that the erosion threshold of the smectite clay increased by more than 10 times as the salinity increased from 0.15 to 1.5 ppt. However, when the salinity was higher than 1.5 ppt, the changes in erosion threshold with increasing salinity were not significant. They further show that the change in the dependency of erosion threshold on salinity coincided with a transformation in the microstructures of the clay. Specifically, at a salinity of less than 1.5 ppt, the clay is from a gel, and as such to erode the gel, the internal bonds of the gel must be broken down. They anticipate that the internal strength of the gelatinous clay network increases with increasing salinity, which is why the erosion threshold increases with increasing salinity. At salinity larger than 1.5 ppt, the clay becomes isolated aggregates. The erosion of these aggregates does not require the internal bonds of the gelatinous network to be broken down, which is why salinity plays a less important role in controlling the erosion threshold. Our recent study also shows that bentonite clay forms a gel at low salinity and becomes liquid-like at high salinity, as shown in **Figure 2**. They further show that given time to consolidate the erosion threshold of the clay doubled due to alterations in its microstructures. However, the impact of consolidation was much smaller than the impact of salinity. These findings emphasize the crucial role of salinity in controlling clay microstructures, rheological properties, and the onset of erosion (**Figure 3**, [23]).

6. Discussion

The studies outlined above offer valuable insights into the aggregation, settling, and erosion behaviors of cohesive sediment-containing clay under diverse salt conditions. However, quantitative predictive models for cohesive sediment transport remain lacking, especially under varying salt conditions. Most of the present studies do not control the mineral composition and the clay content in the sediment, making it difficult to develop quantitative measures of the role of clay on cohesive sediment transport. Moreover, cohesive clay often harbors significant amounts of organic matter and biological components, further complicating predictions of cohesive sediment transport. Therefore, additional research is needed to comprehensively understand how factors, such as clay content, mineral composition, and organic and biological constituents interact to influence cohesive sediment transport under varying salt conditions.

7. Conclusions

This chapter reviews the connection between salt and the movement of cohesive sediment with a focus on clay minerals. First, we discuss the characteristics of clay minerals including their classifications, size distribution, surface area, charges, and EDL. Compared with non-cohesive sand, the large surface area and surface charge of clay minerals contribute to their cohesiveness. The interactions between clay minerals and salt contribute to the dependency of clay and cohesive sediment transport on salt. We also delve into the formation of the layer EDL around clay particles and how this layer influences particle aggregation and dispersion.

In addition to clay minerals, we review studies that explore how salinity affects the transport of sediment. Specifically, we discussed how salinity influences the coagulation, stability, size variations, and settling speed of clay aggregates. Furthermore, we discuss how the SAR and sodicity impact sediment erodibility. These studies emphasize the role of ions such as sodium in regulating soil erodibility across soil types. Furthermore, we discussed a paper that investigated the influence of salinity on the microstructures and erosion threshold of smectite, which reveals a nuanced relationship between salt concentration and clay characteristics. Research has shown that raising the salinity level from 0.15 to 1.5 ppt can increase the erosion threshold of clay by more than 10 times. Similarly, an increase in the SAR from 4 to 73 can lead to a reduction in soil erosion by around 20–24%. These findings underscore the role of salt in regulating sediment movement.

This chapter provides perspectives on how salt interacts with clay particles and their transport in water systems. The various research works summarized in this chapter underscore the important role of salt in controlling clay properties and cohesive sediment transport. A fundamental understanding of particle characteristics, such as clay surface areas, surface charge, and EDL is key to understanding how salt impacts clay transport. More research is needed to improve the fundamental understanding of the impact of salt on the interactions among clay primary particles and the transport of clay and cohesive sediment.

Acknowledgements

The authors of this chapter are supported by the United States National Science Foundation grant EAR 2150796.

Author details

Soukaina Benaich and Judy Q. Yang*
St. Anthony Falls Laboratory, Department of Civil, Environmental, and Geo-Engineering, University of Minnesota Twin Cities, MN, USA

*Address all correspondence to: judyyang@umn.edu

References

[1] Grabowski RC, Droppo IG, Wharton G. Erodibility of cohesive sediment: The importance of sediment properties. Earth-Science Reviews. 2011;**105**:101-120

[2] Huggett JM. Clay minerals. In: Reference Module in Earth Systems and Environmental Sciences. Netherland, Amsterdam: Elsevier; 2015

[3] Nadeau PH. The physical dimensions of fundamental clay particles. Clay Minerals. 1985;**20**:499-514

[4] Sposito G. The Chemistry of Soils. Oxford, United Kingdom: Oxford University Press; 1989

[5] Tan X, Liu F, Hu L, Reed AH, Furukawa Y, Zhang G. Evaluation of the Particle Sizes of Four Clay Minerals. Netherlands: Applied Clay Science; 2017

[6] Okada K. Effect of the crystallinity of kaolinite precursors on the properties of mesoporous silicas. Applied Clay Science. 2008;**41**:10-16

[7] Weil R, Brady N. The Nature and Properties of Soils. 15th ed. New Jersey, USA: Pearson Press; 2017

[8] Liu J, Gaikwad R, Hande A, Das S, Thundat T. Mapping and quantifying surface charges on clay nanoparticles. Langmuir. 2015;**31**:10469-10476

[9] Kumar N, Andersson MP, van den Ende D, Mugele F, Siretanu I. Probing the surface charge on the basal planes of kaolinite particles with high-resolution atomic force microscopy. Langmuir. 2017;**33**:14226-14237

[10] Smith D. Calcium Basics. Available from: https://www.agweb.com/news/crops/crop-production/calcium-basics [Accessed: January 15, 2024]

[11] Bradford SA, Kim H. Implications of cation exchange on clay release and colloid-facilitated transport in porous media. Journal of Environmental Quality. 2010;**39**:2040-2046

[12] Park S-J, Seo M-K. Chapter 1—Intermolecular force. In: Park S-J, Seo M-K, editors. Interface Science and Technology. Vol. 18. Netherland, Amsterdam: Elsevier; 2011. pp. 1-57

[13] Benavides, Arancibia JA. Dielectric response interpretation of shale rocks with low cation exchange capacity. In: Proceedings of the Second International Meeting for Applied Geoscience & Energy; Society of Exploration Geophysicists and American Association of Petroleum Geologists. Houston, Texas: Society of Exploration Geophysicists; 15 Aug 2022. pp. 3568-3570

[14] Liu X, Feng B, Tian R, Li R, Tang Y, Wu L, et al. Electrical double layer interactions between soil colloidal particles: Polarization of water molecule and counterion. Geoderma. 2020;**380**. Article: 114290

[15] Edzwald JK, Upchurch JB, O'Melia CR. Coagulation in estuaries. Environmental Science and Technology. 1974;**8**:58-63

[16] Abolfazli E, Strom K. Salinity impacts on floc size and growth rate with and without natural organic matter. Journal of Geophysical Research Oceans. 2023;**128**(7)

[17] Smoluchowski. Versuch einer mathematischen Theorie der Koagulationskinetik kolloider Lösungen.

Zeitschrift für Physikalische Chemie. 1918;**92**:129-168

[18] Krahl E, Vowinckel B, Ye L, Hsu T-J, Manning AJ. Impact of the salt concentration and biophysical cohesion on the settling behavior of bentonites. Frontiers in Earth Science. 2022;**10**. DOI: 10.3389/feart.2022.1033465

[19] Parwada C, Van Tol J. The nature of soil erosion and possible conservation strategies in Ntabelanga area, Eastern Cape Province, South Africa. Acta Agriculturae Scandinavica. Section B—Soil & Plant Science. 2016;**66**:544-552

[20] Ghadiri H, Hussein J, Dordipour E. The Effect of Soil Salinity and Sodicity on Soil Erodibility, Sediment Transport, and Downstream Water Quality. Brisbane, Australia: ISCO (International Soil Conservation Organisation); 2004

[21] Singer MJ, Janitzky P, Blackard J. The influence of exchangeable sodium percentage on soil erodibility. Soil Science Society of America Journal. 1982;**46**:1239-1243

[22] Carlin JA, Dellapenna TM, Strom K, Noll CJ. The influence of a salt wedge intrusion on fluvial suspended sediment and the implications for sediment transport to the adjacent coastal ocean: A study of the lower Brazos River TX, USA. Marine Geology. 2015;**359**:134-147

[23] Juan JES, Wei WG, Yang JQ. Impact of salinity on the erosion threshold, yield stress, and gelatinous state of a cohesive clay. Journal of Geophysical Research: Earth Surface. 2024;**129**(3). (In print)

Chapter 2

Put the Land Back on the Land: A National Imperative

Daniel E. Canfield Jr, Mina Kiani, Olga Tammeorg, Priit Tammeorg and Timothy J. Canfield

Abstract

The Rivers and Harbors Act of 1899 (RHA1899), the Soil Conservation Act of 1935 (SCA1935), and the Clean Water Act (CWA1972) were pivotal in managing United States (US) waters. RHA1899 provided the Army Corps of Engineers authority to regulate dredge and fill operations. SCA1935 authorized the US Department of Agriculture to combat soil erosion. CWA1972 empowered the US Environmental Protection Agency to regulate point-source pollution. The implementation of the European Soil Charter of 1972 and the 2000 European Water Framework Directive empowered Europe to address similar soil erosion and water quality problems. By 2022, improvements in erosion losses were obtained with soil conservation programs, but continued losses of organic topsoil threatened soil health, human welfare, and water ecosystems. Worldwide losses, estimated at 24 billion tonnes per year, include US and European losses of ~3 billion and 970 million tonnes per year, respectively. Approximately 60% of eroded materials are delivered to rivers and lakes threatening waters. Dredged freshwater sediments, however, have beneficial uses including restoring health of agricultural and forestry lands and water resources. National initiatives involving defense, agriculture, and environmental governmental units are proposed for recycling organic, nutrient-rich aquatic sediments in world-wide *Put the Land Back on the Land* programs.

Keywords: beneficial uses, carbon storage, circular economy, dredging, erosion, lake sediments, national security, nutrients, soil health, water storage

1. Introduction

The world's land and water resources are constantly changing due to natural and anthropogenic forces. Causative agents for these changes are numerous, but a commonality among many identified agents is the influence of erosion. Knowledge of sediment generation, downgradient transport, and deposition have been well described [1, 2] and advances in our understanding of these processes continue as outlined in this book.

Soil erosion and its effects on water quality at catchment scales have become a recent focus for many investigations because of the importance of nonpoint source inputs of nitrogen, phosphorus, and organic carbon to waterways which contribute to the eutrophication of many global waters [3]. Worldwide losses of fertile topsoil have been estimated at 24 billion tonnes per year [4] with United States (US) and European Union (EU) having losses of ~3 billion [5] and 970 million tonnes per year, respectively [6]. By the beginning of the twenty-first century, it was also recognized that 90% of US croplands, due to soil erosion, were losing fertile soil faster than it could be replaced by natural regenerative processes [7, 8]. These erosional losses, besides delivering nutrients and organic carbon to waterways and impacting agricultural production, have reduced reservoir and lake water storage capacity. Deposition of these eroded soils has restricted navigation, made lands adjacent to waterways prone to flooding, and delivered various contaminants (e.g., pesticides and heavy metals) to aquatic systems throughout the world. Approximately 60% of the eroded soil is deposited rivers and lakes [5]. Pimentel [5] concluded that "soil erosion is second only to population growth as the biggest environmental problem the world faces."

The total costs of land degradation to any country alone are difficult to accurately estimate because of the many uncertainties associated with calculating cost estimates [9], but annual global damage from soil erosion has been estimated at $400 billion (USD) with US agricultural losses alone reaching ~$37 billion per year [5]. Removing ~135 million m^3 of sediment annually from EU and United Kingdom (UK) waters [10] is estimated to cost 2.3 billion euros (€) per year (~$2.6 billion USD). In fiscal year 2022, the US Congress allocated $3.44 billion specifically for navigation maintenance and improvements of harbors and inland, coastal, and intracoastal waterways [11].

Removing sediments from waterways is expensive. But to reduce the total costs associated with soil erosion to societies, here we present an approach that we call *Put the Land Back on the Land.* This approach focuses on waterways that are not included in traditional coastal navigation, harbors, and inland and intracoastal waterways. Our proposed approach will assist in mitigating many land and water problems due to soil erosion. We proffer the approach will lead to improved cost/benefit ratios to societies by reincorporating sediments deposited in aquatic systems back onto the watershed landscape.

If soil erosion is the second biggest worldwide environment problem due to its threat to food security [5], our *Put the Land Back on the Land* addresses not only a national imperative of many nations, but an important national security issue for those nations. To understand our contention that now is the time to implement *Put the Land Back on the Land* programs across the world, we discuss four themes:

- Historical Development of the Erosion Problem and Its Management
- Key Steps for Implementing a *Put the Land Back on the Land* Program
- Risk Considerations
- Research for Successful Sediment Recycling

We also identify some of the major socio-political influences on the management of eroded sediments deposited in aquatic systems (Sections 2.1, 2.2). We especially note influences before and after the rise of the environmental movement on the development of regulations imposed by the US and the EU for dredged sediments. Next, we discuss problems with using and interpreting sediment quality guidelines (SQGs) developed for aquatic systems when applied to soils and experiences with sediment reuse in agriculture (Sections 2.3, 2.4). We then discuss key steps needed for implementation of a successful national sediment recycling program (Sections 3.1, 3.2, 3.3, 3.4). Finally, we highlight some risk considerations (Section 4.0), identify research needed for successful sediment recycling (Section 5.0). and conclude that recycling aquatic sediments is inevitable to maintain sustainable productivity of agricultural lands. Given the possible future risks posed by projected climate change and other environmental changes, the loss of vital topsoil, and diminishing agricultural/forestry lands to population growth, now is the time to begin to *Put the Land Back on the Land*!

2. Historical development of the erosion problem and its management

The poet Maya Angelou reportedly stated that you cannot really know where you are going until you know where you have been. The written history on dredging, soil erosion, and their environmental impacts (e.g., eutrophication) are dominated by authors from Europe and the US [12]. This is also true for the beneficial use of dredged material [13]. To understand actions taken by different world governments during specific times, actions must be placed in the context of socio-political norms [14].

2.1 Why soil erosion is a concern?

Humankind through the ages have used water systems for the transportation of people and goods. Waterborne transportation, however, was often hindered by the movement of bottom sediments and the sedimentation of eroded material from upstream watersheds. Humans responded first by removing bottom materials by hand, then by animal power, and finally by use of machines. The breakthrough in sediment removal technology [15, 16] was the development of the suction dredge in the eighteenth century by American Nathaniel Lebby (1857) and *French engineer Henri-Émile Bazin (1867).* Suction dredging facilitated the creation of channels and harbors of sufficient depth for larger and larger ships and became over time one of the commonly used methods in modern dredging [17].

The severity of soil erosion in upstream watersheds depends on geomorphic processes [18]; thus, humankind's responses to soil erosion vary. However, erosion at many worldwide sites became, and still is, a major constraint to agriculture, affecting crop yields and jeopardizing the sustainability of agriculture activities [19, 20]. Land degradation is therefore considered one of the planet's greatest environmental issues [21].

Soil erosion, however, is known to have beneficial agricultural effects downstream from the original fields. Deposition of eroded sediments created the fertile floodplain soils (e.g., the Mississippi River and Nile River basins) and created land (e.g., deltas) that benefited wildlife and humans [22]. *But* regular flooding was required to maintain land fertility and flooding over time was deemed undesirable by humans living alongside flowing waters. Most societies, therefore, established flood control projects (e.g., 1928 Flood Control Act, 70th US Congress, Sess. 1. Ch. 569, enacted May 15, 1928) to prevent damage to infrastructure and provide water storage. Soil erosion mitigation or *Keep the Land on the Land* became the dominant approach to maintain agricultural vitality and there was little consideration of using dredged sediments to maintain soil fertility [23]. Dredged materials were viewed as a waste to be disposed of quickly and cheaply.

Although significant improvements in erosion losses have been obtained in many countries, losses of organic topsoil remain at levels threatening soil health and national welfare [24], raising the question as to why dredged sediments are not routinely recycled back onto agricultural lands. There are of course many reasons put forth, but the toxicity of chemicals in the dredged material became a major concern for the environmental movement in the 1970s [25]. Implementing dredging programs then became problematic because of the necessity to comply with new laws and regulations designed to provide protection for the environment and human health. Besides the scientific and legal issues, there were also social and cultural issues. The number of stakeholders involved increased and each group had varying opinions on what to do. There was angst among stakeholders because there were uncertainties about the impact of a proposed project on other community problems. Most importantly, the economic burden of any dredging project became a confrontational issue. Implementing projects, therefore, became the proverbial *Gordian Knot* and numerous desired dredging projects in the US and Europe reached a standstill.

The United States Environmental Protection Agency (USEPA) and the United States Army Corps of Engineers (USACE) recognized that due to efforts implemented under the Clean Water Act (CWA1972: 3 U.S.C. §1251 et seq. 1972) that both sediment and water quality have improved to a point where toxicity criteria are often achieved. Dredged materials from many US water bodies were now deemed "clean enough" to be recommended for beneficial uses [26, 27]. For example, one identified use after appropriate evaluations (physical, chemical, and biological) was applying dredged material to agricultural and forestry lands to reclaim and conserve topsoil. Similar conclusions were also reached in Europe [28].

2.2 Past efforts and challenges

Waterways have served as transportation routes throughout history. People also dug channels to irrigate crops. Maintaining waterways for commercial and national defense became a priority for many nations in the nineteenth century as ships of war became larger and more goods and people were transported by larger ships. In the case of the US, the USACE was assigned the mission of constructing and maintaining reliable, navigation systems. USACE and military organizations in Europe were the only organizations in the nineteenth century that had the engineers needed to improve and develop new dredging technology as well as the ability to address the economic and national security imperatives of their country.

Until the 1970s, dredged materials were viewed as beneficial in Europe and the US. They were used to fill coastal and inland marshlands, to create land for building sites, roads, and rails to cross lowlands, as well as for agriculture. For many countries, wetlands were considered an obstacle to development. This belief was best illustrated in the US when Congress passed the Swamps Land Act of 1850 (Thirty-First Congress, Chapter 84) that provided the mechanism for transferring title of federally owned wetlands to states that would drain the land for productive agricultural uses. The nutrient-rich soil under wetlands was especially attractive as agricultural land once drained.

In the twentieth century, the 1930s Dust Bowl in the US focused attention on soil erosion (Public Law 74-76, Statute 163). The Dust Bowl was viewed as a national disaster because the uprooting of topsoil posed a *national security* threat [29]. Soil erosion, however, was treated as a natural resource management issue, not a national defense issue. The US Congress established the Soil Conservation Service (SCS) to combat soil erosion and "to preserve natural resources, control floods, prevent impairment of reservoirs, and maintain the navigability of rivers and harbors, protect public health, public lands and relieve unemployment." Consequently, SCS initiated a program for building water bodies across the US to trap sediments eroding from the land and increase water storage for agricultural and forestry operations.

The problem of soil erosion became severe enough in Europe to be part of the environmental agenda of the EU due to its impacts on food production, drinking water quality, eutrophication, and ecosystem services [6]. Small reservoirs became the preferred infrastructure for use against intensive runoff processes [30]. As in the US, European small reservoirs increased in number to provide greater water storage and trapping of sediments [31].

By the 1970s, Earth Day and the rising environmental movement established a new "Environmental Paradigm" [32, 33]. Concurrently, a new social movement emphasizing the importance of "Deep Ecology" emerged, a phrase originated in 1972 by Norwegian philosopher Arne Naess [34, 35]. This social change led to the promulgation of many new laws, regulations, and policies in both Europe and the US as well as worldwide [36, 37], with the two pieces of iconic legislation being the CWA1972 and the 2000 European Water Framework Directive (Directive 2000/60/EC).

The CWA1972 was passed in response to the discharge of hazardous substances into rivers, the loss of wetlands by dredging and in-filling, and nutrient enrichment, which were becoming national environmental concerns. The 1977 CWA [Section 404(f)], however, exempted discharges associated with normal agricultural and forestry activities. By 1987, many of the discharges of toxic materials from point sources had been removed or plans to remove them established. Parties interested in eutrophication issues expanded the focus from point source controls to include nonpoint sources of pollution (Section 319 of CWA). Although agriculture was recognized as a major source of nonpoint source nutrients, exemptions for agriculture remained in place for the US. The agricultural exemptions recognized that American agriculture fulfilled the public need for supplying abundant and affordable food and fiber. Other nations also had similar exemptions for agriculture.

By the 1970s, many world scientists also began to focus on the eutrophication of lakes [38]. The Clean Lakes Program (CWA Section 314) directed the USEPA to develop better methods for removing silt, aquatic growth, and other obstructions in lakes. USEPA was also directed to create silt traps and devices to prevent sediment deposits in lakes and demonstrate the benefits of using dredged material for land reclamation. Subsequently, many international conventions (London Convention of 1972, Barcelona Convention of 1976, Helsinki Convention of 1992, and OSPAR Convention of 1992) encouraged the exploration of alternative uses for dredged materials. Such efforts, however, were stymied due to a lack of funding and concerns about using contaminated sediments. USEPA came to view dredging and other palliative approaches (e.g., aquatic weed control) as "*short-term*" solutions for eutrophication problems and used regulation 40 CFR Part 35.1650-2 to disallow funding for many desired dredging projects and other lake management palliative methods.

Accepting that dredged materials could have beneficial uses was also hindered because agronomic units of government were focused on *Keep the Land on the Land* programs, that is, erosion control. Environmental governmental units, which often blamed agriculture and dredging for the loss of wetlands [39], began to focus on sediment contamination issues. This concern about contaminants in sediments, coupled with limited data in the early stages, fostered the belief that dredged sediments should always be treated as a waste!

The management approach for dredged material, therefore, became to place sediments into confined disposal facilities (CDF) or dispose materials in the open water of large lakes (e.g., the US and Canadian Great Lakes) or the oceans. Despite the holding capacity of CDFs becoming limited and the dumping of materials in open waters being restricted in the twenty-first century, legislative limitations developed in the late twentieth century hindered the recycling of dredged sediments, especially in the EU [40].

2.3 Sediment contamination

Research on contaminated underwater sediments evolved slowly in the latter part of the twentieth century. One of the most important regulatory tools for aquatic sediment management has been the development of sediment quality guidelines (SQGs). Guidelines were based on the hypothesis that a chemical concentration level could be identified in sediments below which aquatic life would not be harmed. The Threshold Effect Limit (TEL) hypothesis was first proposed by the USEPA and USACE in the early 1970s to assess the disposal of dredged material at sea [41, 42]. The contaminated sediments issue, however, was not specifically addressed in the laws of the US until the Water Resources Development Act of 1992 (WRDA).

Since the 1980s, numerous SQGs have been established by different organizations and countries with each SQG incorporating different criteria to account for the varied conditions in which sediment contamination occurs. Consequently, identifying the concentration threshold for chemicals below which aquatic life was not harmed remained problematic [43] and only guidelines, not numerical legal standards, exist for most nations.

The development of SQGs led to the use of both the TEL and Probable Effect Limit (PEL). As noted previously, the TEL is the sediment contamination concentration at which a toxic response was first observed for benthic test organisms. The PEL was the concentration at which a great percentage of the benthic population showed a toxic response. Concentrations above the PEL were deemed problematic. When a chemical concentration was between the TEL and PEL, professional judgment was to be used to assess the consequences of sediment contamination.

Because of the uncertainty factors (UF, generally not reported) inherent in different SQG approaches, nations are still developing SQGs in the twenty-first century. In South America, SQGs that are meaningful for the unique biological conditions of tropical and high-altitude environments are being developed [44]. However, the more significant issue to consider for agricultural interests is that SQGs have been developed for aquatic systems, not agricultural lands. If the sediment is not going to be re-used in a wet system (e.g., wetland restoration), SQGs are of limited value for dry land application. Consequently, various guidelines for agricultural soil health have been developed, and there is an ongoing need to define contaminant risk levels for agricultural soils. Most agricultural TELs are greater than aquatic TELs (**Table 1**).

In the case of heavy metals, the background concentrations at a particular geographic location need to be considered. In highly mineralized soils, existing aquatic SQGs might be exceeded, resulting in unnecessary restrictions on the recycling of dredged sediments. Spreading dried dredged material in thin layers across a large agricultural landscape links to the historical practice "Dilution is the Solution to Pollution." Dredged sediments and the agricultural lands proposed for deposition should, therefore, be tested for contaminants. Each nation has established toxicity levels for various contaminants that are likely to elicit toxic effects. If the spreading process is done so that the final contaminant concentration is below the toxicity levels known to

Substance	Sediment			Soil[3]		Compost[4]
	TEL[1]	Lowest effect Level [2]	Severe effect level[2]	TEL	Lower guideline value	Agricultural use
Metals						
Arsenic (As)	5.9	6	33	5	50	23
Cadmium (Cd)	0.6	0.6	9	1	10	1
Cobalt (Co)				20	100	
Chromium (Cr)	37.3	26	110	100	200	70
Copper (Cu)	35.7	16	110	100	150	150
Iron (Fe)						
Manganese (Mn)						
Nickel (Ni)	18	16	50	50	100	60
Lead (Pb)	35	31	110	60	200	120
Antimony (Sb)		2	25	2	10	
Vanadium (V)				100	150	
Zinc (Zn)	123	120	270	200	250	500
Mercury (Hg)	0.17			0.5	2	0.7
PAHs and PCBs						
Anthracene	50	85.3	1100	1000	5000	
2-Methylnaphthalene		70	670			
Acenaphthene	10	16	500			
Acenaphthylene	10	44	640			
Benzo(a)anthracene	70	261	1600	1000	5000	
Benzo(a)pyrene	90	430	1600	200	2000	
benzo(b)fluoranthene	70			1000	5000	
benzo(ghi)perylene						
benzo(k)fluoranthene	60					
Chrysene	110	384	2800			
Dibenz(a,h) anthracene		63,4	260			
Fluoranthene	110	600	5100	1000	5000	
Fluorene	20	19	540			
Henanthrene						
Naphthalene	30	160	2100	1000	5000	
Indeno(1,2,3-cd) pyrene						
Phenanthrene	90	240	1500	1000	5000	
Pyrene	150	665	2600			

Substance	Sediment			Soil[3]		Compost[4]
	TEL[1]	Lowest effect Level[2]	Severe effect level[2]	TEL	Lower guideline value	Agricultural use
Total PAH	870	4022	44,792	15,000	30,000	10,000
Total PCBs	34	22.7	180	100	500	100

[1]TEL, threshold effect level; Represents the concentration below which adverse effects on sediment-dwelling organisms are unlikely to be observed regarding metals [45] and organic contaminants [46].
[2]Screening criteria for sediments in New York State. The Lowest Effect Level indicates a level of sediment contamination that can be tolerated by the majority of benthic organisms, but still causes toxicity to a few species. The Severe Effect Level indicates the concentration at which pronounced disturbance of the sediment dwelling community can be expected. A sediment is considered contaminated if either criterion is exceeded. If both criteria are exceeded, the sediment is considered to be severely impacted. If only the Lowest Effect Level criterion is exceeded, the impact is considered moderate [47, 48].
[3]The Finnish legislation sets concentration levels by each hazardous element to identify soil contamination and remediation needs. Threshold value is equally applicable for all sites, and it indicates the need for further assessment of the area. In areas where background concentration is higher than the threshold value, background concentration is regarded as the assessment threshold. The second concentration level is the so-called "guideline value." If this is exceeded, the area has a contamination level which presents ecological or health risks. Different guideline values are set for industrial and transport areas (higher guideline value) and for all other land uses (lower guideline value) [49].
[4]Maximum limit values of heavy metal concentration in compost "class. A" which is suitable for agriculture in Europe [50].

Table 1.
Screening criteria for metals (mg kg^{-1}), polycyclic aromatic hydrocarbon (PAHs; μg kg^{-1} dry weight), and total polychlorinated biphenyls (PCBs; μg kg^{-1} dry weight) in sediment, soil, and compost.

cause toxic effects in organisms, there should be no environmental impact. The United States Department of Agriculture (USDA) developed simple equations to identify the maximum amount of material that can be applied to a given amount of land [51].

Land application experience with materials previously considered waste has been gained by spraying or spreading biosolids (sewage sludge) onto the land surface. Both the EU and US emphasize the beneficial effect of recycling nutrient rich biosolids, while seeking to reduce risks from pathogens and other contaminants [52, 53]. Injecting of biosolids beneath the land surface or incorporating the biosolids into the soil by discing the soil can serve to either condition the soil or fertilize crops and vegetation grown in the soil. Landfarming is also a "Dilution is the Solution to Pollution" biotreatment technology that uses available agricultural methods to enhance biodegradation of organic contaminants. The Dilution is *A* solution to pollution approach should not be summarily discounted because it has been branded with a bad connotation in the past. Sediments, once spread in thin layers and mixed into the underlying soil, permit natural attenuation processes to degrade and immobilize many contaminants [54].

While the presence of chemical contaminants has occupied the attention of many scientists during the latter part of the twentieth century, one of the other issues that was a major concern was the eutrophication of aquatic systems from point source and nonpoint source nutrient inputs [38, 55].

2.4 Sustainability (eutrophication, the circular economy, soil health, and plant growth)

Concerns regarding eutrophication and the approaches to use to mitigate adverse effects of nutrient enrichment dominated world-wide research efforts in the late twentieth century [55]. Once Vollenweider [56] introduced the concept of nutrient loading to the scientific community, governmental environmental agencies began to focus on limiting point source nutrient inputs, especially phosphorus, to waterways with many successes [57]. However, it was realized that lakes were very complex and often responded differently to point source and nonpoint source nutrient control strategies [58]. This was especially true for shallow (<3 m), productive lakes where the internal supply of nutrients was great [59–61]. In many systems, internal nutrients, especially phosphorus, were mobilized (released) into the water column as sediments became anoxic [62, 63] or wind mixing brought nutrient-rich sediments into the water preventing improvements in water quality [61, 64].

Conversion of natural landscapes to agricultural and urban uses occurred throughout the world with population growth and changing climate. Human activities, however, often increased erosion, sediment transport, and sedimentation in aquatic systems. Erosion rates can be 100 to 1000 times greater than the geological erosion rate [65]. The implications for eutrophication are considered well known [66, 67]. Changes in agricultural operations [68] and the construction of wetlands [39, 69–71] have been some of the measures employed to try to control nonpoint source nutrient inputs. Best management practices (BMPs) have proven to be largely inadequate for restoring eutrophic lakes; thus, additional measures will be needed to reduce internal nutrient supplies [61, 72].

Explanations for the inadequate BMP responses have centered on legacy P [73, 74] and N [75] inputs from the waterbody's watershed stymying management programs [76–78]. Soil phosphorus buildup due to past management practices (Legacy P) is often considered a root cause for eutrophication. Legacy nutrients also exist in lake and river sediments [64, 79]. Missimer et al. [64] concluded that despite extensive efforts to control external nutrient loads into Florida's Lake Okeechobee, the frequency of algal blooms will continue until the sediments containing legacy nutrients are removed [61, 63].

Khare et al. concluded that the most cost-effective approach for achieving an average flow-weighted total phosphorus (TP) concentration of 40 μg/L at Lake Okeechobee's inflows would require an estimated $4.26 billion [70]. This approach involves the implementing of nutrient management BMPs, Dispersed Water Management programs, as well as the construction of ~200 km^2 of Wetland Restoration and Stormwater Treatment Areas (STAs). The total costs and the need for such a large wetland area may be daunting for all but the most affluent nations and those with the capacity to convert a great quantity of agricultural/forestry lands to wetlands. Whether or not wetlands could reduce nonpoint source nutrient inputs sufficiently to lower Lake Okeechobee's TP remains speculative and does not address the Legacy P problem.

Federal and state scientists estimated that fully dredging Lake Okeechobee would cost at least $1 billion, but <$4.26 billion for the establishment of BMPs, and restored wetlands and the construction of STAs [80]. Dredging would remove decades of Legacy P helping reduce in-lake TP and algal bloom frequencies [61, 64]. Deepening Okeechobee would increase water storage and provide water needed for agriculture and the growing population of south Florida [14]. Reducing the need to raise lake levels would also decrease detrimental ecological damage to Okeechobee's marshes [14, 81]. Past efforts to reduce nonpoint nutrient inputs to Lake Okeechobee have failed to meet legally established goals [61, 64], but dredging will provide benefits even if nutrient inputs are not decreased [82].

Many lake restoration experts, however, now support removing bottom sediments to promote the concept of sustainable lake restoration [83]. Sustainable lake restoration can improve water quality, while delivering numerable socio-economic benefits. It also melds well with the Circular Economy concept developed in Europe [84, 85]. The Circular Economy Action Plan (CEAP) was accepted by the European Commission in March 2020. The concept, however, has not been readily adopted in other nations because of limited documentation of any successes. There are also cultural barriers, a lack of consumer interest and awareness, as well as a hesitant company culture and concerns by policy makers [86]. Removal of nutrients from lake ecosystems for their recovery and re-use (i.e., circular economy) should be prioritized [63].

Because the concept of reusing sediment aligned with circular economy principles and coincided with growing interests in global sustainability, research on the impact of reused sediment on soil and plant growth increased in the early twenty-first century. Studies demonstrated that there were significant benefits for applying aquatic sediments to agricultural lands [40, 87–90]. Benefits included improved topsoil organic matter content, a greater amount of available water and nutrients for plants, and improved crop yields [40, 87–91]. As a soil amendment, some recycled sediments also improved soil cation exchange capacity [87, 92], elevated soil pH [93, 94], and decreased soil bulk density [87, 95].

Application of recycled lake sediment has been tested on vegetables and fruits [87, 93, 96], field crops like cereals and grasses [96–100], and ornamentals [101, 102] with positive effects on yield response. In general, studies where the recycled sediments were tested against control soils that were adequately fertilized similar growth resulted. As an example, Woodard [95] reported that forage crops (alfalfa and big bluestem) grown in a South Dakota lake's sediment had similar or slightly higher shoot dry matters and higher relative N and P uptakes compared with those plants grown in silty clay loam soil. Ebbs et al. [93] also suggested that reclaimed slightly alkaline silty loam sediment from Lake Peoria (Illinois) could be utilized for vegetable production.

Crop yield benefits have been most notable in nutrient-deficient soils [87, 92, 95, 97]. A recent promising example of sediment recycling for soil fertilization comes from Lake Mustijärv (Estonia), where sediment removal and recycling led to increased growth and nutrient uptake of grass mixture, demonstrating the potential of sediment as a valuable resource in both short and longer terms [98, 99, 103]. Similar approaches have also been developed at Lake Ormstrup (Denmark) to reduce internal phosphorus loading and establish links to circular economy principles [104]. These efforts highlighted the growing interest in sediment recycling as a sustainable agricultural practice for closing the human-related nutrient cycle.

However, not all sediment recycling experiments have yielded positive results. Reduction in crop yields, due to soil compaction of loamy sand soil, occurred when sediment from Besko Reservoir (Poland) was added [105]. Chechlo Reservoir (Poland) sediment with a 2.7-fold lower carbon content and poor P and K fertility also reduced crop yields [106]. Also, sediment from constructed Finland wetlands had very low P availability for plants [107, 108], probably because of great amounts of P-binding compounds in sediments. Sediment originating from diverse sources (e.g., agriculture, industry, or urban areas) will have varying soil properties and may contain pollutants such as heavy metals and organic pollutants [40]. Given these variations in sediments quality and the associated ecological risks, it is essential to conduct appropriate evaluations of the sediments and to characterize their compositions to appropriately apply amendments and treatments when warranted before utilizing them in agricultural lands.

3. Key steps for implementing a *Put the Land back on the Land* Program

In the twenty-first century, "Sustainability" emerged as a scientific goal to find ways to use natural resources in a manner to guarantee their continuity and the continuity of humanity itself [109]. *Put the Land Back on the Land* is an approach to sustainability for each nation's agricultural and forestry lands. However, education will be an important component if the approach is to be successful for dredged materials that have been viewed as a waste (often a hazardous waste) since the 1970s. Also, the number of stakeholders involved has increased and each has varying opinions on what to do. Conflict resolution will become important to avoid the proverbial *Gordian Knot*.

3.1 Prioritizations

Every nation has issues that must be prioritized, with national defense being a top priority for all nations. A key component of all national defense strategies is food security [110–112]. Global population increases generally lead to a reduction of farmland due to increasing urbanization. Urbanization further limits global food production capacity [113, 114]. Losses of organic topsoil also threaten soil health and national welfare [24, 29]. Consequently, the USDA established a soil health division in 2015. Feeding more people with less farmland or less productive farmland, therefore, requires technological changes and the development of appropriate public policies to ensure vital food security within the global security system.

Regenerative farming has garnered increased attention in the twenty-first century as a technological change that offers promise to resolve some issues related to agricultural lands [115, 116]. This approach, however, maintains the same soil management philosophy of *Keep the Land on the Land*. While retaining the *Keep the Land on the Land* approach is still valid, it is not sufficient alone to address erosion and loss of topsoil. To address the many intertwining issues associated with erosion (e.g., the shallowing of waterways and eutrophication), a holistic *Put the Land Back on the Land* freshwater sediment recycling program and needed infrastructure enhancements must be developed in addition to the current *Keep the Land on the Land* approach to enhance and maintain the productivity of agricultural and forestry lands through the remainder of the twenty-first century.

3.2 Funding

Limited local governmental budgets, as well as limited local sponsor budgets, discourage the removal of sediments from waterbodies and the use of more costly beneficial use alternatives, even if those alternatives are more environmentally sustainable [117]. Alternatively, many port dredging projects are funded at the national level. The Food & Global Security Network (NETWORK) convened a meeting of 22 experts (military minds, NGO leaders, scientists, and practical farmers) to discuss the critical importance of soil health [118]. The NETWORK called on national governments to formally recognize healthy soil as a strategic asset and critical for maintaining national security.

Approximately 877 billion dollars (USD) were spent on US national defense in 2022 and 17 billion dollars (USD) were spent on Army Civil Works studies and projects (Public Law 117-58), to maintain existing infrastructure, repair damage, and dredge channels in response to floods and coastal storms. NETWORK experts urged each nation's defense departments to work with their agricultural and environmental counterparts to jointly oversee delivery of increased food sovereignty as well as the regeneration of soil function. Each agency can no longer work just within their own legislatively mandated silo if environmental sustainability is to be achieved. This philosophical shift emphasizes that holistic management can be a more efficient approach for managing soil erosion problems and restoring the vitality of agricultural lands.

In the US, a 2% annual increase in the nation's defense budget would provide adequate funding (~17 billion annually) to support the operation of a national *Put the Land Back on the Land* program for freshwater systems and develop the infrastructure needed to support such a program. With adequate funding, *Put the Land Back on the Land* initiatives can aid in closing an important environmental loop that would limit nutrient loss from agricultural lands. Each nation can adjust funding to meet their national need.

3.3 Contacting potential stakeholders

First, the *Put the Land Back on the Land* approach is workable only if supported by landowners willing to accept reclaimed sediments. These landowners will include agricultural/forestry landowners, or other landowners desiring organic sediments for soil augmentation. Due to past erosion control programs, many if not most agricultural landowners have small water bodies on their land to trap sediments. Initial efforts should focus on these waters as the landowners recognize the trapped sediments came from their land. Governmental agricultural organizations should test these sediments for soil quality to demonstrate how testing will also be done on sediments from other water bodies. This knowledge should help to alleviate contamination concerns by not only the agricultural landowner but other stakeholders agreeing to accept reclaimed sediments.

Involving landowners directly is key to helping with an understanding of the benefits of putting the material back on their land. Some landowners with waterbodies on their properties, however, may not initially want governmental employees or contractors on their land. If the "Do It Myself" attitude becomes the limiting factor, equipment such as agitator pumps and soaker hoses can be provided to those landowners for delivery of sediments to the surrounding lands. When placing large amounts of water on the land would interfere with intended agricultural/forestry operations or contribute to new soil erosion, parabolic wedge-wire filters or a vertical geotube dewatering tower (**Figure 1**) could be utilized. These devices offer a simple economical approach for both small and large landowners needing to remove sediments from their waterbodies. The parabolic wedgewire or the vertical dewatering systems not only permit a return of the water to the water body but provide a form of sediment that can be spread on the land using available (owned or rented) farm equipment.

For landowners not wanting to work on their waters, but receiving truckable sediments from distant waters, centralized processing facilities can be developed and scaled to the size of the waterbody having sediments removed. From these facilities, dry material of appropriate quantity can be delivered, per agreement with the landowner, to the agricultural lands without disrupting normal agricultural operations.

One encouraging example of the ways to engage landowners is a decision-making tool for Brazilian farmers who commonly use sediments from close-by lakes and ponds that dry out regularly, making the dredging more feasible. The app (https://resed-cientista-chefe-v1.streamlit.app/, App Developer- Brennda Bezerra Braga Ph.D) facilitates farmers to calculate the economically feasible distance for transportation of the sediments based on the properties of soils and sediments and the costs of alternative, mineral fertilizer.

The advantage of first working directly with landowners is that it will build their support for accepting beneficial material from other water sources. A list of willing

Figure 1.
Parabolic rundown screen (a) and vertical geotube tower (b) removing dredge material at farm sites. MuckVac™ sediment processing unit (c) and application tractor (d) used by AGUACULTURE of Clearwater, FL.

landowners will facilitate determination of the available acreage for receiving aquatic sediment enhancements. Successful sediment removal will, by removing decades of legacy nutrients deposited in the water bodies, reestablishes the sediment trapping ability of waterways, thus reducing the delivery of erodible material downstream. The cooperative work efforts will open the opportunity for conservation staff to ensure effective erosion management strategies (e.g., filter strips and sediment retention ponds) are in place. More importantly, the landowners, especially the agricultural/ forestry owners, may then become part of the solution for mitigating ongoing eutrophication by receiving potentially more nutrients that are released by their lands.

Many freshwater water bodies located downstream of agricultural/forestry lands have accumulated sediments that restrict intended uses of the water bodies. Each nation's environmental organization should work with non-landowning stakeholders to prioritize waters for refurbishment and the establishment of sustainable lake restoration programs [83]. These stakeholders need to be informed the *Put the Land Back on the Land* program is not a program to drain wetlands or remove base materials from a water body to increase water storage. The approach for sediment removal is analogous to vacuuming the nutrient-rich organic matter off the bottom to provide material to replenish lost topsoil.

The non-landowning stakeholders will be informed removed sediments will be tested for contaminants and results reported to them and the soil recipients to ensure soil health on lands proposed for disposal. In Florida, the Florida Department of Environmental Protection (FDEP) discourages agricultural landowners from placing material on their land if nutrients can re-enter waterways to contribute to nonpoint source nutrient loads. All stakeholders and regulatory agencies must acknowledge that multiple decades of legacy nutrients will be removed from these waterways. Even though there is the possibility of future inputs of some of these nutrients coming back into these waterways, the net reduction of nutrients should outweigh what re-enters from the reclaimed sediments.

Another concern of landowners and non-landowners is the loss of water during sediment removal operations. When there is sufficient water, where loss of water does not become a major stakeholder concern, and nearby agricultural or forest lands are reachable (~1.6 km), a vacuum system (MuckVac™) that ensures only the free-floating, unconsolidated muck is extracted from the lakebed has been used in South Florida. The extracted sedimentary material is conveyed *via* a diesel pump and reinforced lay-flat hoses, to a specialized Application Unit. If needed, removed material can be stored in mobile frac tanks until applied to the land. The Application Unit is an agricultural-grade tractor that pulls the lay-flat hose through the application field (**Figure 1**). The discharge of the material is meticulously calibrated to agronomic rates. To reach a distant site (>1.6 km), additional pumps are used (AGUACULTURE, www.agcutech.com, Clearwater, FL).

For many waters, the stakeholders do not want water removed from their water body and regulatory agencies often apply water quality standards to the returning dredge water. In 2000, there was another breakthrough in dredging technology, the Genesis Rapid Dewatering System™ (RDS) by Genesis Water (**Figure 2**). The RDS system permitted rapid dewatering of dredged sediments. Separation of solids and water from the dredged materials helped mitigate the size of needed disposal areas and thus reduce the overall project footprint [119]. Because the RDS system is scalable, waters in urbanized areas could be dredged. Dredged materials were also repurposed after physical, chemical, and biological characterization [119]. Less contaminated or uncontaminated sediments generally are suitable for commercial or beneficial reuse without treatment [120]. The RDS system also treats the return water to meet local water quality standards. Returning water to the dredged system has a tremendous economic impact for project developers.

Since the introduction of RDS, improvements in the system have been made *by Revive Water Technologies of Bloomington, Illinois*. The system now provides a continuous 1:1 operating ratio of dredging to dewatering, timely production of stackable and truckable solids, immediate release of clear water discharge, and mobility with minimal energy and maintenance. Besides removing legacy nutrients from the return water, the system also greatly reduces the concentrations of various water quality contaminants from these same return waters (**Table 2**).

Having the dewatered sediments truckable is a great advantage as uncontaminated material can be delivered to more distant recipient lands. As dry sediments, large amounts of material could be transported by rail to distant agricultural lands.

Figure 2.
A 15 m^3/min Genesis Rapid Dewatering System™ (a), clear return water (b), and truckable dry material (c) as operated by Revive Water Technologies of Bloomington, Illinois.

Transportation by rail would also permit contaminated sediments to be shipped to remote decontamination centers where treatments could be applied to make material suitable for agricultural uses.

3.4 Legislative laws, regulations, policies, and potential bottlenecks

Each nation has a myriad of laws, regulations, and policies governing dredging and disposal of removed materials. Typically, there is little meshing because the laws, regulations, and policies were developed for each regulatory agency staying within its narrowly defined legislative mission. Sometimes agency missions overlap

Parameters	Influent concentration	Revive RDS concentration	Reduction percentage
Arsenic	160 µg/L	3.3 µg/L	97.9%
Copper	790 µg/L	2.3 µg/L	99.7%
Iron	290,000 µg/L	170 µg/L	99.9%
Lead	9100 µg/L	8.2 µg/L	99.9%
Nickel	250 µg/L	2.3 µg/L	99.1%
Mercury	4.0 µg/L	0.062 µg/L	98.5%
Total nitrate	16.7 mg/L	11.35 mg/L	31.9%
Phosphorous	4500 µg/L	48 ppb	98.9%
Total suspended solids	150,000 mg/L	15 mg/L	99.9%

Table 2.
Sawgrass Lake—demonstration of rapid dewatering technology. U.S. Lake Dredging Project: RDS Contaminant Reduction (modified table re-produced with permission from Author) [121].

but there is a tendency for each agency to only cite the laws, regulations, and policies used by their agency. When broad-based program like *Put the Land Back on the Land* is advanced, confusion often follows. For example, Finland and the Czech Republic permit direct reuse of sediments dredged from inner water bodies on agricultural soils if contaminant concentrations meet their national legislation thresholds [40]. Other European nations do not, and current EU actions place limitations on the application of sediments in agriculture [40].

Dredged sediments, recognized as natural resources, pose challenges in terms of management and disposal, contingent on factors like sediment type, pollution levels, and proximity to disposal facilities near dredging areas [40]. The diverseness of these environmental factors has made it difficult to formulate universal sediment management regulations. Consequently, the EU has indirectly addressed dredged material management by applying existing directives like the EU Water Framework Directive, EU Waste Directive, and EU Directive on protected areas, rather than a dedicated framework directive [40]. The EU Water Framework Directive checks water quality, but contamination limits for sediments vary by member state. The Waste Directive classifies sediments as either hazardous or non-hazardous waste, potentially allowing reuse (receiving codes in the European Waste Catalogue 170,505 (hazardous waste) or 170,506 (non-hazardous waste)). The Groundwater Directive oversees dredging activities, with a focus on groundwater quality and flood control. The European Landfill Directive does not cover using dredged sludge for fertilization or depositing non-hazardous sludges near water bodies, and the Habitat and Birds Directives safeguards biodiversity and controls dredging near nature protection sites.

The lack of specificity in existing European legislative guidance causes uncertainty about the legality of the *Put Land Back on the Land* approaches and may hinder or stop its implementation. The use of recycled sediments on agricultural lands as well as their use as fertilizers therefore needs greater legislative clarity. The laws, regulations, and policies also need to be reviewed to facilitate greater agency cooperation for achieving each nation's goal to achieve environmental sustainability. In the US, USACE and USEPA have worked closely on sediment removal and beneficial use efforts since the end of the twentieth century and bringing closer cooperation/joint efforts with the USDA-National Resource Conservation Service will greatly enhance future management efforts.

4. Risk considerations

Each society will have to confront issues where there are pros and cons to be considered. For policy makers, decisions are often influenced by uncertainty. Here, a distinction must be made between generic uncertainty as a human (i.e., not knowing or understanding risk to oneself or family if there is any presence of a contaminant at all) and the use of scientific uncertainty factors in the law and toxicology. When a water quality standard is placed in legislative law, the published number is associated with an uncertainty factor of 10, 100, or 1000 depending on the quality and type of toxicological study conducted [122].

When extrapolating from valid experimental results, a 10-fold factor is applied [123]. This is done in studies involving prolonged exposure to average, healthy humans. This factor is intended to account for the variation in sensitivity among the members of the human population. An additional 10-fold factor is used when extrapolating from valid results of long-term studies on experimental animals (the results of studies of human exposure are not available or are inadequate). This factor is intended to account for the uncertainty involved in extrapolating from animal data to humans. Finally, an additional 10-fold factor is used when extrapolating from less than chronic results on experimental animals and there are no useful long-term human data [124].

When dealing with recycled sediments or soil health, and in the absence of specific data sufficient to support the setting of legislatively authorized standards, regulatory agencies use guidelines when setting regulations and policies. Their use is generally based on professional judgment. The Society of Environmental Toxicology and Chemistry (SETAC, offices in North America and Europe) recommended aquatic sediment management decisions should be based on site-specific information generated to evaluate the predictive ability of SQGs at a site of interest [43]. Sediment assessment frameworks for aquatic systems as for soil assessment frameworks, therefore, must be flexible for different management purposes. Policy makers must use their judgment to determine the greater good when dealing with recycled sediments because guidelines may not adequately account for the combined impact of various contaminants when they are present together, as opposed to situations involving only one type of contaminant [125]. While it is well recognized that the operational approach to environmental risk management strategies for sediments is to reduce the potential risk of contaminants, as total elimination of environmental risk from contaminants is typically not feasible or achievable, the lack of total risk elimination of contaminants from recovered sediments should not necessarily preclude the reuse and reincorporation of these sediments back on the landscape. Evaluations of potential risks from multiple contaminants can be evaluated, and potential concentration level limits can be developed to provide the desired level of protection that would allow these sediments to be safely reincorporated onto the landscape for beneficial uses.

As a matter of practicality, when only a TEL for an aquatic system is available, multiplying that number by an uncertainty factor of 10 will most likely approach a guideline derived for soil. If that number is associated with another uncertainty factor of 10, there is still a good safety factor for soil. While this approach would be technically sound for sediments from many waters, there will be a limited number of waters where the sediments might have to be classified as non-usable to avoid risks. However, the understanding of risk continues to evolve and numbers between a TEL and PEL are in a gray zone. Professional judgments will still be needed to interpret risk factors until better toxicological data are derived [122–124]. Whenever soil-based screening level guidances are available, they should be examined for use when considering land-based applications of dredged sediments. In the absence of such soil-based screening level guidances, then use the above stated TEL with uncertainty factor approach.

5. Research for successful sediment recycling

The implementation of sediment recycling programs such as *Put the Land Back on the Land* necessitates a continued comprehensive and multidisciplinary approach. To address the intricate challenges associated with sediment management, it is imperative to merge insights from diverse fields including environmental science, agriculture, engineering, and law. Such an approach will be needed to continue developing scalable, cost-effective, sediment removal technologies that can adapt to individual environmental systems.

Past research initiatives have focused on understanding the impact of sediment recycling on both soil environmental conditions and waterbody ecosystem, in addition to assessing potential benefits for plant production. The sediment excavation and reuse initiatives in Lake Mustijärv and the ongoing efforts in Lake Ormstrup emphasize the importance of merging limnology, agriculture, and engineering expertise [98, 99, 103, 104]. In addition, engineering research has contributed innovations such as the MuckVac™ system and the Genesis Rapid Dewatering System™ (RDS), streamlining the extraction and reuse processes. These technologies showcase the collaborative efforts of engineers and environmental scientists in creating efficient and sustainable sediment management tools. Moreover, efforts need to be made to address economic impacts. The economic loss due to the decline of land productivity (on-site effects) caused by water erosion is estimated at about 1.25 billion € per year in the EU [126]. Considering economic principles, agricultural advancements, and techniques for sediment removal provides a holistic perspective on *Put the Land Back on the Land*.

6. Conclusions

There are certain inevitabilities that all societies on earth will have to deal with regularly. Sediment erosion will continue and there will be a loss of organic topsoil and nutrients. Sedimentation will result in the shallowing of waterways, and eutrophication of aquatic systems will continue. Throughout the twentieth century, knowledge regarding how to mitigate these problems advanced considerably and improvements have been made. Techniques for removing sediments from waterways to support commercial and/or recreational uses advanced to the point where removed sediments can now be rapidly dewatered and the water returned to the targeted waterway safely. Beneficial uses for dredged material have also been identified, but continued education will become paramount for success [109].

Agriculture fulfills the vitally important public need for all nations in supplying abundant and affordable food and fiber. The loss of organic topsoil and the diminishing productivity of agricultural lands will, therefore, become the premier concern for many nations in the remainder of twenty-first century as food security *IS* national security.

American President Dwight D. Eisenhower [127] during his 1961 farewell speech spoke about the military-industrial complex and discussed the people's expectation that political leaders must find essential agreement on issues of great moment, the wise resolution of which will better shape the future of a Nation. He also noted that all proposals must be weighed in the light of a broader consideration: the need to maintain balance in and among national programs—balance between the private and the public economy, balance between cost and hoped for advantage—balance between the clearly necessary and the comfortably desirable; balance between essential requirements as a nation and the duties imposed by the nation upon the individual; and balance between actions of the moment and the national welfare of the future. Good judgment seeks balance and progress; lack of it eventually finds imbalance and frustration. We submit the holistic *Put the Land Back on the Land* program meets these criteria and avoids the impulse to live only for today [128] without considering environmental sustainability for future generations.

Sufficient knowledge regarding the recycling of dredged materials now exists and in the words of American President Franklin Delano Roosevelt "We have nothing to fear but fear itself" [129]. The management goal for all dredged materials should, therefore, be to use removed material beneficially unless chemically unsuited to remain in the environment. The soil management philosophy of *Keep the Land on the Land* (i.e., soil conservation) has made important contributions to the worldwide conservation effort and needs to continue in the future. The *Keep the Land on the Land* management approach, however, failed to close an important environmental loop for nutrients, ultimately threatening environmental sustainability. Recycling aquatic sediments to agricultural lands is inevitable. The time is now that Nations should adopt and implement the *Put the Land Back on the Land* approach to foster sustainability and resiliency of both the aquatic and terrestrial resources that are imperatives to not only their individual national interests, but more broadly the collective Global Planetary interests that benefit all peoples.

Acknowledgements

We thank our past and present scientific and lake management colleagues for obtaining the vast amount of information needed to finally help us see the many linkages between the different scientific disciplines and environmental problems. We especially appreciate the multiple conversations regarding the issues discussed in this paper that led to the recognition *Put the Land Back on the Land* programs are a national imperative for all nations.

We thank Brian Dyson and Ken Forshay of the USEPA Office of Research and Development (ORD) and James Grundy of the USEPA Office of Land and Emergency Management (OLEM) for their critical review comments which helped improve this chapter. We also thank editor Andrew J. Manning for his editorial comments. Olga Tammeorg was supported by Maa- ja Vesitekniikan tuki ry (project nr 4644) and Estonian Research Council (grant PRG 1167).

Conflict of interest

The authors declare no conflict of interest. The views expressed in this chapter are those of the author(s) and do not necessarily represent the views or the policies of the USEPA. Any mention of trade names, manufacturers, or products does not imply an endorsement by the United States Government or the USEPA.

Author details

Daniel E. Canfield Jr[1*], Mina Kiani[2,3], Olga Tammeorg[3,4], Priit Tammeorg[3] and Timothy J. Canfield[5]

1 School of Forest, Fisheries, and Geomatics Sciences, Institute of Food and Agricultural Sciences, University of Florida, Gainesville, Florida, USA

2 Natural Resources Institute Finland, Finland

3 Department of Agricultural Sciences, University of Helsinki, Helsinki, Finland

4 Estonian University of Life Sciences, Tartu, Estonia

5 Office or Research and Development, U.S. Environmental Protection Agency, Ada, Oklahoma, USA

*Address all correspondence to: decan@ufl.edu

References

[1] Rose CW. Erosion and sedimentation. In: Bonell M, Hufschmidt MM, Gladwell JS, editors. Hydrology and Water Management in the Humid Tropics: Hydrological Research Issues and Strategies for Water Management. Cambridge: Cambridge University Press; 1993. pp. 301-343. DOI: 10.1017/CBO9780511564468

[2] Haan CT, Barfield BJ, Hayes JC. Design Hydrology and Sedimentology for Small Catchments. Cambridge: Academic Press; 1994. 588 p

[3] Merritt WS, Letcher RA, Jakeman AJ. A review of erosion and sediment transport models. Environmental Modelling & Software. 2003;**18**:761-799. DOI: 10.1016/S1364-8152(03)00078-1

[4] UNCCD [United Nations Convention to Combat Desertification]. Land and Soil in the Context of a Green Economy for Sustainable Development, Food Security and Poverty Eradication. Bonn. Ger. Available from: https://catalogue.unccd.int/850_Rio_6_pages_english.pdf: UNCCD [United Nations Convention to Combat Desertification]; 2011 [Accessed: November 11, 2023]

[5] Pimentel D. Soil erosion: A food and environmental threat. Environment, Development and Sustainability. 2006;**8**:119-137. DOI: 10.1007/s10668-005-1262-8

[6] Panagos P, Borrelli P, Poesen J, Ballabio C, Lugato E, Meusburger K, et al. The new assessment of soil loss by water erosion in Europe. Environmental Science & Policy. 2015;**54**:438-447. DOI: 10.1016/j.envsci.2015.08.012

[7] USDA. Changes in Average Annual Soil Erosion by Water on Cropland and CRP Land, 1992-1997. Washington D.C.: United States Department of Agriculture, Natural Resources Conservation Service; 2000

[8] USDA. Changes in Average Annual Soil Erosion by Wind on Cropland and CRP Land, 1992-1997. Washington D.C: United States Department of Agriculture, Natural Resources Conservation Service; 2000

[9] Graves AR, Morris J, Deeks LK, Rickson RJ, Kibblewhite MG, Harris JA, et al. The total costs of soil degradation in England and Wales. Ecological Economics. 2015;**119**:399-413. DOI: 10.1016/j.ecolecon.2015.07.026

[10] Panagos P, Matthews F, Patault E, De Michele C, Quaranta E, Bezak N, et al. Understanding the cost of soil erosion: An assessment of the sediment removal costs from the reservoirs of the European union. Journal of Cleaner Production. 2023;**434**:140183. DOI: 10.1016/j.jclepro.2023.140183

[11] Normand AE, Carter NT. Army Corps of Engineers: FY 2022 Appropriations. CRS IF11846. Washington D.C.: Congressional Research Service; 24 Jun 2022. Available from: https://crsreports.congress.gov/product/pdf/IF/IF1184 [Accessed: December 22, 2023]

[12] Bhagowati B, Ahamad KU. A review on lake eutrophication dynamics and recent developments in lake modeling. Ecohydrology & Hydrobiology. 2019;**19**(1):155-166. DOI: 10.1016/j.ecohyd.2018.03.002

[13] Solanki P, Jain B, Hu X, Sancheti G. A review of beneficial use and management of dredged material.

Waste. 2023;**1**:815-840. DOI: 10.3390/waste1030048

[14] Canfield DE Jr, Bachmann RW, Hoyer MV. Socio-Political Influences Affecting the Management & Restoration of Okeechobeeland: Lake Okeechobee and the Greater Everglades. Gainesville FL. 131 p. Available from: https://lakewatch.ifas.ufl.edu/media/lakewatchifasufledu/extension/bibliography/Socio-politial-influences----Okeechobeeland-copy.pdf: Florida LAKEWATCH, School Forest, Fisheries and Geomatics Sciences, UF/IFAS; 2023 [Accessed: October 20, 2023]

[15] Manders D. Research and Development in the US Army Corps of Engineers: Improving the Common Stock of Knowledge. St. Louis, MO: US Army Corps of Engineers; 2011. Available from: https://apps.dtic.mil/sti/pdfs/ADA581204.pdf [Accessed: November 4, 2023]

[16] Lebby MM, Nathaniel H. Coverage August 22, 1816–February 11, 1880. Institute for Southern Studies. Columbia, SC: University of South Carolina; 2016. . Available from: https://www.scencyclopedia.org/sce/entries/lebby-nathaniel-h/ [Accessed: November 11, 2023]

[17] Hayes DF. Dredging 101: Fundamentals of dredging. WEDA [Western Dredging Association] Temecula. CA. Available from: https://www.westerndredging.org/phocadownload/Dredging_101/WEDA%20Dredging%20101-%20Session%201.pdf [Accessed: December 28, 2023]

[18] Walling D, Webb B. Erosion and sediment yield: Global and regional perspectives. In: Proceedings of the Exeter Symposium. Vol. 236. Exeter, UK: IAHS; 1996. pp. 3-19. Available from: https://www.researchgate.net/publication/255578368_Erosion_and_Sediment_Yield_A_Global_Overiew [Accessed: November 28, 2023]

[19] Pimentel D, Harvey C, Resosudarmo P, Sinclair K, Kurz D, McNair M, et al. Environmental and economic costs of soil erosion and conservation benefits. Science. 1995;**267**(5201):1117-1123. DOI: 10.1126/science.267.5201.1117

[20] Ahmad NSBN, Mustafa FB, Yusoff SYM, Didams G. A systematic review of soil erosion control-practices on the agricultural land in Asia. International Soil and Water Conservation Research. 2020;**8**:103-115. DOI: 10.1016/j.iswcr.2020.04.001

[21] IPBES [Intergovernmental Science-Policy Platform on Biodiversity and Ecosystem Services]. The IPBES assessment report on land degradation and restoration. In: Montanarella L, Scholes R, Brainich A, editors. Secretariat of the Intergovernmental Science-Policy Platform on Biodiversity and Ecosystem Services. Bonn, DE: IPBES; 2018. DOI: 10.5281/zenodo.3237392

[22] Warburton P. River basin management: The Nile and Mississippi. Geofile. 2002;**416**. 4 pages. Available from: https://www.jkgeography.com/uploads/1/0/8/4/108433405/mississippi_and_the_nile.pdf [Accessed: December 28, 2023]

[23] Dotterweich M. The history of human-induced soil erosion: Geomorphic legacies, early descriptions and research, and the development of soil conservation—A global synopsis. Geomorphology. 2013;**201**:1-34. DOI: 10.1016/j.geomorph.2013.07.021

[24] Toor G, Yang Y, Das S, Dorsey S, Felton G. Chapter four-soil health in

agricultural ecosystems: Current status and future perspectives. In: Sparks D, editor. Advances in Agronomy. Vol. 168. Cambridge, Massachusetts: Academic Press; 2021. pp. 157-201. DOI: 10.1016/bs.agron.2021.02.004

[25] Vogt C, Hartman G. Dredging practices and environmental considerations. In: Meyers RA, editor. Encyclopedia of Sustainability Science and Technology. New York, NY: Springer; 2012. DOI: 10.1007/978-1-4419-0851-3_438

[26] Williams K, Margeson K, Paczuski S, Auker K, Hoffman J. Beneficial Use of Dredge Material Decision Tool: RESES Project Report. Washington D.C.: U.S. Environmental Protection Agency; 2020. Available from: https://cfpub.epa.gov/si/si_public_record_report.cfm?Lab=CCTE&dirEntryId=352253 [Accessed: November 11, 2023]

[27] USACE [U.S. Army Corps of Engineers]. Beneficial Uses of Dredged Sediment. Vicksburg, Mississippi: US Army Engineer Research and Development Center Dredging Operations Technical Support Program; 2023. Available from: https://budm.el.erdc.dren.mil [Accessed: November 11, 2023]

[28] USAR [Using Sediments as A Resource]. Using Sediments as Resource: Sediment Recycling Strategy. Interreg 2 Seas Programme. Rotterdam, Netherlands: European Union; 2023. Available from: www.schielandendekrimpenerwaard.nl.documents [Accessed: November 26, 2023]

[29] Bennett HH, Chapline WR. Soil Erosion a National Menace. Washington D.C., USA. Circular NO. 33. Available from: https://archive.org/details/soilerosionnatio33benn/page/n1/mode/2up: United States Department of Agriculture; 1928 [Accessed: November 26, 2023]

[30] Selman P. What do we mean by sustainable landscape? Sustainability: Science, Practice and Policy. 2008;**4**(2):23-28. DOI: 10.1080/15487733.2008.11908019

[31] Honek D, Šulc Michalková M, Smetanová A, Sočuvka V, Velísková Y, Karásek P, et al. Estimating sedimentation rates in small reservoirs - suitable approaches for local municipalities in Central Europe. Journal of Environmental Management. 2020;**261**:109958. DOI: 10.1016/j.jenvman.2019.109958

[32] Carson R. Silent Spring. Boston: Houghton Mifflin Company; 1962. 154 p. Available from: https://archive.org/details/fp_Silent_Spring-Rachel_Carson-1962/mode/2up [Accessed: October 20, 2023]

[33] White L. The historical roots of our ecological crisis. Science. 1967;**155**:1203-1207. DOI: 10.1126/science.155.3767.1203

[34] Duvall B, Sessions G. Deep Ecology: Living as if Nature Mattered. Salt Lake City: Peregrine Smith Books; 1985; 267 p

[35] Sponsel LE. Spiritual Ecology, A Quiet Revolution. Santa Barbara: Praeger; 2012. 285 p

[36] Fisher E. The history of environmental law. In: Environmental Law: A Very Short Introduction. Oxford: Oxford Academic; 2017. 144 p. DOI: 10.1093/actrade/9780198794189.003.0004

[37] Deters H. European environmental policy at 50: Five decades of escaping decision traps. Environmental Policy and Governance. 2019;**29**:315-325. DOI: 10.1002/eet.1855

[38] NAS [National Academy of Sciences]. Eutrophication: Causes,

Consequences, Correctives. In: Proceedings of a Symposium. Washington DC: NAS [National Academy of Sciences]; 1969; 661 p. DOI: 10.17226/20256

[39] Allord TD, Dahl TE. History of wetlands in the conterminous United States. In: Fretwell JD, Williams JS, Redman PJ, editors. National Summary on Wetland Resources. Washington, D.C.: US Geological Survey Water-Supply Paper 2425; 1996. pp. 19-26. DOI: 10.3133/wsp2425

[40] Renella G. Recycling and reuse of sediments in agriculture: Where is the problem? Sustainability. 2021;**13**(4):1648. DOI: 10.3390/su13041648

[41] USEPA/USACE. Technical committee on criteria for dredged and fill material. In: Ecological Evaluation of Proposed Discharge of Dredged Material into Ocean Waters, Implementation Manual for Section 103 of Public Law 92-532 (Marine Protection, Research, and Sanctuaries Act of 1972), July 1977 (Second Printing April 1978). Vickburg, MS. Available from: https://apps.dtic.mil/sti/tr/pdf/ADA143776.pdf: Environmental Effects Laboratory, U. S. Army Engineer Waterways Experiment Station; 1973 [Accessed: November 28, 2023]

[42] Engler RM, Long ER, Swartz RC, Di Toro DM, Ingersoll CG, Burgess RM, et al. Chronology of the development of sediment quality assessment methods in North America. In: Wenning R, Batley G, Ingersoll C, Moore D, editors. Use of Sediment Quality Guidelines and Related Tools for the Assessment of Contaminated Sediments. Pensacola: SETAC Press; 2005. pp. 311-343. 783 p

[43] Wenning RJ, Ingersoll CG. Summary of the SETAC Pellston Workshop on Use of Sediment Quality Guidelines and Related Tools for the Assessment of Contaminated Sediments. Pensacola: Society of Environmental Toxicology and Chemistry (SETAC); 2002. 48 p. Available from: http://www.setac.org/files/SQGSummary.pdf [Accessed: November 26, 2023]

[44] Nascimento IA. Key issues on sediment quality assessment: A review of South America research with emphasis on Brazil. Aquatic Ecosystem Health & Management. 2007;**10**:9-22. DOI: 10.1080/14634980701209189

[45] MacDonald DD, Ingersoll CG, Berger TA. Development and evaluation of consensus-based sediment quality guidelines for freshwater ecosystems. Archives of Environmental Contamination and Toxicology. 2000;**39**:20-31. DOI: 10.1007/s002440010075

[46] Swartz RC. Consensus sediment quality guidelines for polycyclic aromatic hydrocarbon mixtures. Environmental Toxicology and Chemistry: An International Journal. 1999;**18**(4):780-787. DOI: 10.1002/etc.5620180426

[47] NYSDEC [New York State Department of Environmental Conservation]. Technical Guidance for Screening Contaminated Sediments. Albany: NYSDEC; 1999. 39 p. Available from: https://lmpublicsearch.lm.doe.gov/SiteDocs/SW-A-006230.pdf [Accessed: November 28, 2023]

[48] O'Connor TP. The sediment quality guideline, ERL, is not a chemical concentration at the threshold of sediment toxicity. Marine Pollution Bulletin. 2004;**49**:383-385. DOI: 10.1016/j.marpolbul.2004.06.024

[49] MEF [Ministry of the Environment]. Government Decree on the Assessment of Soil Contamination and Remediation Needs. Finland: MEF; 2007. p. 214.

Available from: https://finlex.fi/en/laki/kaannokset/2007/en20070214.pdf [Accessed: November 28, 2023]

[50] Amlinger F, Pollak M, Favoino E. Heavy Metals and Organic Compounds from Wastes Used as Organic Fertilisers. Report to the European Commission, DG-Environment. Report/Project ENV.A.2./ETU/2001/0024. 2004. European Commission, DG, Environment, Brussels, Belgium. Available from: https://ec.europa.eu/environment/pdf/waste/compost/hm_finalreport.pdf [Accessed: November 5, 2023]

[51] USDA [United States Department of Agriculture]. Heavy Metal Soil Contamination. Soil Quality – Urban Technical Note No. 3. Auburn, AL: Natural Resources Conservation Service Soil Quality Institute; 2000. Available from: https://directives.sc.egov.usda.gov/18583.wba [Accessed: November 28, 2023]

[52] Iranpour R, Cox HHJ, Kearney RJ, Clark JH, Pincince AB, Daigger GT. Regulations for biosolids land application in U.S. and European Union. Journal of Residuals Science & Technology. 2004;**1**(4):209-222. Available from: https://d3pcsg2wjq9izr.cloudfront.net/files/9256/articles/5809/Regulations.pdf [Accessed: November 11, 2023]

[53] USEPA [U.S. Environmental Protection Agency]. Basic Information about Biosolids. Washington, D.C.: USEPA; 2023. Available from: https://www.epa.gov/biosolids/basic-information-about-biosolids [Accessed: November 28, 2023]

[54] Balland-Bolou-Bi C, Brondeau F, Dung Jusselme M. Can natural attenuation be considered as an effective solution for soil remediation? In: Environmental Sciences. London, United Kingdom: IntechOpen; 2023. DOI: 10.5772/intechopen.108304

[55] Farley M. Eutrophication in fresh waters: An international review. In: Bengtsson L, Herschy RW, Fairbridge RW, editors. Encyclopedia of Lakes and Reservoirs. Encyclopedia of Earth Sciences Series. Dordrecht, NL: Springer; 2012. pp. 258-270. DOI: 10.1007/978-1-4020-4410-6_79

[56] Vollenweider RA. Advances in defining critical loading levels for phosphorus in lake eutrophication. Memorie Dell Instituto Italiano D'Idrobiologia. 1976;**33**:53-83

[57] Schindler DW, Carpenter SR, Chapra SC, Hecky RE, Orihel DM. Reducing phosphorus to curb lake eutrophication is a success. Environmental Science & Technology. 2016;**50**(17):8923-8929. DOI: 10.1021/acs.est.6b02204

[58] Schindler DW. Recent advances in the understanding and management of eutrophication. Limnology and Oceanography. 2006;**51**(1):356-363. DOI: 10.4319/lo.2006.51.1_part_2.0356

[59] Søndergaard M, Jensen JP, Jeppesen E. Role of sediment and internal loading of phosphorus in shallow lakes. Hydrobiologia. 2003;**506-509**(1-3):135-145. DOI: 10.1023/B:HYDR.0000008611.12704.dd

[60] Cooke GD. History of eutrophic lake rehabilitation in North America with arguments for including social sciences in the paradigm. Lake and Reservoir Management. 2007;**23**(4):323-329. DOI: 10.1080/07438140709354021

[61] Canfield DE Jr, Bachmann RW, Hoyer MV. Restoration of lake Okeechobee: Mission impossible. Lake and Reservoir Management.

2021;**37**(1):95-111. DOI: 10.1080/10402381.2020.1839607

[62] Nürnberg GK. Assessing internal phosphorus load – Problems to be solved. Lake and Reservoir Management. 2009;**25**:419-432. DOI: 10.1080/00357520903458848

[63] Tammeorg O, Nürnberg GK, Tõnno I, Toom L, Nõges P. Spatio-temporal variations in sediment phosphorus dynamics in a large shallow lake: Mechanisms and impacts of redox-related internal phosphorus loading. Science of the Total Environment. 2024;**907**:168044. DOI: 10.1016/j.scitotenv.2023.168044

[64] Missimer TM, Thomas S, Rosen BH. Legacy phosphorus in lake Okeechobee (Florida, USA) sediments: A review and new perspective. Water. 2020;**13**(1):39. DOI: 10.3390/w13010039

[65] Julien PY. Erosion and Sedimentation. 2nd ed. Cambridge: Cambridge University Press; 2010. 365 p. Available from: https://archive.org/details/erosionsedimenta0000juli/page/n5/mode/2up [Accessed: November 28, 2023]

[66] Soranno PA, Hubler SL, Carpenter SR, Lathrop RC. Phosphorus loads to surface waters: A simple model to account for spatial pattern of land use. Ecological Applications. 1996;**6**:865-878. DOI: 10.2307/2269490

[67] Jeppesen E, Sondergaard M, Kronvang B, Jensen JP, Svendsen LM, Lauridsen TL. Lake and catchment management in Denmark. Hydrobiologia. 1999;**395**(396):419-432. DOI: 10.1023/A:1017071602549

[68] Withers PJA, Neal C, Jarvie HP, Doody DG. Agriculture and eutrophication: Where do we go from here? Sustainability. 2014;**6**(9):5853-5875. DOI: 10.3390/su6095853

[69] Mitsch WJ, Gosselink JG. Wetlands. 2nd ed. New York: John Wiley and Sons; 1993; 722 p. DOI: 10.2134/jeq1994.00472425002300050040x

[70] Khare Y, Naja GM, Stainback GA, Martinez CJ, Paudel R, Van Lent T. A phased assessment of restoration alternatives to achieve phosphorus water quality targets for Lake Okeechobee, Florida, USA. Water. 2019;**11**(2):327. DOI: 10.3390/w11020327

[71] Ferreira CSS, Kašanin-Grubin M, Solomun MK, Sushkova S, Minkina T, Zhao W, et al. Wetlands as nature-based solutions for water management in different environments. Current Opinion in Environmental Science & Health. 2023;**33**:100476. DOI: 10.1016/j.coesh.2023.100476

[72] Osgood RA. Inadequacy of best management practices for restoring eutrophic lakes in the United States: Guidance for policy and practice. Inland Waters. 2017;**7**(4):401-407. DOI: 10.1080/20442041.2017.1368881

[73] Kronvang B, Bechmann M, Lundekvam H, Behrendt H, Rubæk GH, Schoumans OF, et al. Phosphorus losses from agricultural areas in river basins. Journal Environmental Quality. 2005;**34**:2129-2144. DOI: 10.2134/jeq2004.0439

[74] Sharpley A, Jarvie HP, Buda A, May L, Spears B, Kleinman P. Phosphorus legacy: Overcoming the effects of past management practices to mitigate future water quality impairment. Journal of Environmental Quality. 2013;**42**:1308-1326. DOI: 10.2134/jeq2013.03.0098

[75] Basu NB, Van Meter KJ, Byrnes DK, Van Cappellen P, Brouwer R,

Jacobsen BH, et al. Managing nitrogen legacies to accelerate water quality improvement. Nature Geoscience. 2022;**15**(2):97-105. DOI: 10.1038/s41561-021-00889-9

[76] Reddy KR, Newman S, Osborne TZ, White JR, Fitz HC. Phosphorous cycling in the greater Everglades ecosystem: Legacy phosphorous implications for management and restoration. Critical Reviews in Environmental Science and Technology. 2011;**41**(sup1):149-186. DOI: 10.1080/10643389.2010.530932

[77] Motew M, Chen X, Booth EG, Carpenter SR, Pinkas SS, Zipper C, et al. The influence of legacy P on lake water quality in a midwestern agricultural watershed. Ecosystems. 2017;**20**(8):1468-1482. DOI: 10.1007/s10021-017-0125-0

[78] Khare YP, Paudel R, Wiederholt R, Abiy AZ, Van Lent T, Davis SE, et al. Watershed response to legacy phosphorus and best management practices in an impacted agricultural watershed in Florida, U.S.A. Land. 2021;**10**(9):977. DOI: 10.3390/land10090977

[79] Koch S, Rosewig EI, Lennartz B. Legacy phosphorus in sediments of lowland waterways. Environments. 2023;**10**(3):43. DOI: 10.3390/environments10030043

[80] GAO [United States General Accounting Office]. Comprehensive everglades restoration plan: Additional Water Quality Projects May Be Needed and Could Increase Costs. Testimony Before the Subcommittee on Transportation and Infrastructure, Committee on Environment and PublicWorks, U.S. Senate. GAO/T-RCED-00-297. 2000. Available from: https://apps.dtic.mil/sti/tr/pdf/ADA383560.pdf [Accessed: November 29, 2023]

[81] Gray PN, Farrell CJ, Kraus ML, Gromnicki AH, editors. Lake Okeechobee: A Synthesis of Information and Recommendations for its Restoration. Miami (FL): Audubon of Florida. Available from: https://fl.audubon.org/sites/default/files/audubon_stateofthelake_2006.pdf; 2005 [Accessed: December 28, 2023]

[82] Canfield DE Jr, Bachmann RW, Hoyer MV. The Florida experience: Time for a revitalized and restructured clean lakes program. Lake and Reservoir Management. 2023;**39**:191-212

[83] Tammeorg O, Chorus I, Spears B, Nõges P, Nürnberg GK, Tammeorg P, et al. Sustainable lake restoration: From challenges to solutions. WIREs Water. 2024;**11**:e1689. DOI: 10.1002/wat2.1689

[84] Kirchherr J, Reike D, Hekkert M. Conceptualizing the circular economy: An analysis of 114 definitions. Resources, Conservation and Recycling. 2017;**127**:221-232. DOI: 10.1016/j.resconrec.2017.09.005

[85] Kara S, Hauschild M, Sutherland J, McAloone T. Closed-loop systems to circular economy: A pathway to environmental sustainability? CIRP Annals. 2022;**71**(2):505-528. DOI: 10.1016/j.cirp.2022.05.008

[86] Kirchherr J, Piscicelli L, Bour R, Kostense-Smit E, Muller J, Huibrechtse-Truijens A, et al. Barriers to the circular economy: Evidence from the European Union (EU). Ecological Economics. 2018;**150**:264-272. DOI: 10.1016/j.ecolecon.2018.04.028

[87] Canet R, Chaves C, Pomares F, Albiach R. Agricultural use of sediments from the Albufera Lake (eastern Spain). Agriculture, Ecosystems & Environment. 2003;**95**(1):29-36. DOI: 10.1016/S0167-8809(02)00171-8

[88] Sigua GC, Holtkamp ML, Coman SW. Assessing the efficacy of dredged materials from lake Panasoffkee, Florida: Implication to environment and agriculture part 1: Soil and environmental quality aspect. Environmental Science and Pollution Research. 2004;**11**:321-326. DOI: 10.1007/BF02979646

[89] Sigua GC, Holtkamp ML, Coleman SW. Assessing the efficacy of dredged materials from lake Panasoffkee, Florida: Implication to environment and agriculture: Part 2: Pasture establishment and forage productivity. Environmental Science and Pollution Research. 2004;**11**:394-399. DOI: 10.1007/BF02979659

[90] Brigham RD, Pelini S, Xu Z, Vázquez-Ortega A. Assessing the effects of lake-dredged sediments on soil health: Agricultural and environmental implications for Northwestern Ohio. Journal of Environmental Quality. 2021;**50**(2):494-503. DOI: 10.1002/jeq2.20199

[91] Leue M, Lang F. Recycling soil nutrients by using channel deposits as fertilizers? Nutrient Cycling in Agroecosystems. 2012;**93**:75-88. DOI: 10.1007/s10705-012-9501-5

[92] Darmody RG, Diaz DR. Dredged Sediment: Application as an Agricultural Amendment on Sandy Soils. Champaign, Illinois: TR Series (Illinois Sustainable Technology Center); TR-066; 2017. Available from: https://www.frtr.gov/matrix/documents/Dredged-Material-Processing-Technologies/2017-Dredged-Sediment-Application-as-an-Agricultural-Amendment-on-Sandy-Soil.pdf [Accessed: November 28, 2023]

[93] Ebbs S, Talbott J, Sankaran R. Cultivation of garden vegetables in Peoria pool sediments from the Illinois river: A case study in trace element accumulation and dietary exposures. Environment International. 2006;**32**(6):766-774. DOI: 10.1016/j.envint.2006.03.013

[94] Baran A, Tarnawski M, Urbaniak M. An assessment of bottom sediment as a source of plant nutrients and an agent for improving soil properties. Environmental Engineering and Management Journal. 2019;**18**(8):1647-1656. DOI: 10.30638/EEMJ.2019.155

[95] Woodard HJ. Plant growth on soils mixed with dredged lake sediment. Journal of Environmental Science & Health Part A. 1999;**34**(6):1229-1252. DOI: 10.1080/10934529909376893

[96] Tozzi F, Renella G, Macci C, Masciandaro G, Gonnelli C, Colzi I, et al. Agronomic performance and food safety of strawberry cultivated on a remediated sediment. Science of the Total Environment. 2021;**796**:148803. DOI: 10.1016/j.scitotenv.2021.148803

[97] Edesi L, Kangor T, Loide V, Vettik R, Tamm I, Kennedy HJ, et al. Effects of lake sediment on soil chemical composition, dehydrogenase activity and grain yield and quality in organic oats and spring barley succession. Agronomy Research. 2020;**8**(3):2022-2032. DOI: 10.15159/AR.20

[98] Kiani M, Raave H, Simojoki A, Tammeorg O, Tammeorg P. Recycling lake sediment to agriculture: Effects on plant growth, nutrient availability, and leaching. Science of the Total Environment. 2021;**753**:141984. DOI: 10.1016/j.scitotenv.2020.141984

[99] Kiani M, Zrim J, Simojoki A, Tammeorg O, Penttinen P, Markkanen T, et al. Recycling eutrophic lake sediments into grass production: A four-year field experiment on agronomical and environmental implications. Science of the Total Environment. 2023;**870**:161881. DOI: 10.2139/ssrn.4238459

[100] Gmitrowicz-Iwan J, Ligęza S, Pranagal J, Smal H, Wójcikowska-Kapusta A. Improving acidic sandy soil properties for plant growth with dam reservoir sediments in the face of soaring fertiliser prices. Soil and Tillage Research. 2023;**234**:105843. DOI: 10.1016/j.still.2023.105843

[101] Mattei P, Gnesini A, Gonnelli C, Marraccini C, Masciandaro G, Macci C, et al. Phytoremediated marine sediments as suitable peat-free growing media for production of red robin photinia (Photinia x fraseri). Chemosphere. 2018;**201**:595-602. DOI: 10.1016/j.chemosphere.2018.02.172

[102] Ugolini F, Mariotti B, Maltoni A, Tani A, Salbitano F, Izquierdo CG, et al. A tree from waste: Decontaminated dredged sediments for growing forest tree seedlings. Journal of Environmental Management. 2018;**211**:269-277. DOI: 10.1016/j.jenvman.2018.01.059

[103] Kiani M, Tammeorg P, Niemistö J, Simojoki A, Tammeorg O. Internal phosphorus loading in a small shallow lake: Response after sediment removal. Science of the Total Environment. 2020;**725**:138279. DOI: 10.1016/j.scitotenv.2020.138279

[104] Haasler S, Christensen ML, Reitzel K. Synthetic and biopolymers for lake restoration–an evaluation of flocculation mechanism and dewatering performance. Journal of Environmental Management. 2023;**331**:117199. DOI: 10.1016/j.jenvman.2022.117199

[105] Baran A, Jasiewicz C, Tarnawski M. Effect of bottom sediment supplement to light soil on the content and uptake of macroelements by maize. Ecological Chemistry and Engineering. A. 2012;**19**(8):863-871. DOI: 10.1051/e3sconf/20130104008

[106] Koniarz T, Baran A, Tarnawski M. Agronomic and environmental quality assessment of growing media based on bottom sediment. Journal of Soils and Sediments. 2022;**22**(4):1355-1367. DOI: 10.1007/s11368-022-03173-4

[107] Laakso J, Uusitalo R, Leppänen J, Yli-Halla M. Sediment from agricultural constructed wetland immobilizes soil phosphorus. Journal of Environmental Quality. 2017;**46**(2):356-363. DOI: 10.2134/jeq2016.09.0336

[108] Laakso J, Uusitalo R, Heikkinen J, Yli-Halla M. Phosphorus in agricultural constructed wetland sediment is sparingly plant-available. Journal of Plant Nutrition and Soil Science. 2017;**180**(5):554-562. DOI: 10.1002/jpln.201700062

[109] Rull V. The most important application of science. EMBRO Reports. 2014;**15**(9):919-922. DOI: 10.15252/embr.201438848

[110] Bassou A. The Importance of Agricultural Policy in Terms of Security, Defense and Sovereignty. Policy Brief PB-16/12. Rabat, Morocco: OCP Policy Center; 2016. Available from https://www.policycenter.ma/publications/importance-agricultural-policy-terms-security-defense-and-sovereignty [Accessed: October 20, 2023]

[111] Arsil T, Syaiful A. The role of the Ministry of Defense in preparing strategic logistics reserve for facing the National Food Emergency. Journal of Social and Political Sciences. 2021;**4**(2):136-147. DOI: 10.31014/aior.1991.04.02.282

[112] Sarjito A. The role of the Ministry of Defense in ensuring food security for state defense. Journal of Governance. 2022;7(3):573-583. DOI: 10.31506/jog.v7i3.14607

[113] Vliet JV, Eitelberg DA, Verburg PH. A global analysis of land take in cropland areas and production displacement from urbanization. Global Environmental Change. 2017;**43**:107-115. DOI: 10.1016/j.gloenvcha.2017.02.001

[114] Xie Y, Hunter M, Sorensen A, Nogeire-McRae T, Murphy R, Suraci JP, et al. Lark. U.S. farmland under threat of urbanization: Future development scenarios to 2040. Land. 2023;**12**:574. DOI: 10.3390/land12030574

[115] Carlisle L, De Wit MM, DeLonge MS, Calo A, Getz C, Ory J, et al. Securing the future of US agriculture: The case for investing in new entry sustainable farmers. Elementa: Science of the Anthropocene. 2019;**7**:17. DOI: 10.1525/elementa.356

[116] Carlisle L. Healing Grounds: Climate, Justice, and the Deep Roots of Regenerative Farming. United States: Island Press; 2022; 200 p

[117] Searcy Bell K, Boyd B, Goetz S, Hayes D, Magar V, Suedel B. Overcoming barriers to beneficial use of dredged material in the US. WEDA Journal of Dredging. 2021;**19**(2):20-42. Available from: https://westerndredging.org/phocadownload/2021_Virtual/Proceedings/2A-4.pdf [Accessed: November 10, 2023]

[118] Network [Food & Global Security Network]. Soil Health: A Security Threat Profile. Dorset, UK: Farmwell; 2021. Available from: https://www.foodandsecurity.net/_files/ugd/0f4d79_7291a83f21fa4c558927a767997825f8.pdf [Accessed: October 5, 2023]

[119] Hodges M, Emmons S, Cascioli K. High-speed dewatering and stabilization of contaminated sediments, Columbus (Ohio), USA. Paper K-25. In: Durell GS, Foote EA, editors. Conference Chairs. Remediation of Contaminated Sediments—Fifth International Conference. 2-5 February 2009. Columbus, OH. Jacksonville, Florida. Available from: www.battelle.org/sedimentsconhttps://www.mcilvainecompany.com/Decision_Tree/subscriber/Tree/DescriptionTextLinks/bob%27s%20confidentiality%20agree.pdf: Battelle Memorial Institute; 2009 [Accessed: November 26, 2023]

[120] Olin-Estes TJ, Palermo MR. Recovery of dredged material for beneficial use: The future role of physical separation processes. Journal of Hazardous Materials. 2001;**85**:39-51. DOI: 10.1016/S0304-3894(01)00220-5

[121] Almaleh LJ, Roppelt G. Sawgrass Lake – Demonstration of rapid dewatering technology. In: Presented at Western Dredging Association (WEDA) Gulf Coast Chapter Meeting, November 18, 2015 in New Orleans, Louisiana. Temecula, CA: Western Dredging Association; 2015

[122] Gibson MC, deMonsabert SM, Orme-Zavaleta J. Comparison of noncancer risk assessment approaches for use in deriving drinking water criteria. Regulatory Toxicology and Pharmacology. 1997;**26**(3):243-256. DOI: 10.1006/rtph.1997.1150

[123] Degen GH, Seiler J, Bentley P. Toxicology in transition. In: Proceedings of the 1994 EUROTOX Congress Meeting Held in Basel, Switzerland, August 21-24, 1994. New York, NY: Springer; 1995. DOI: 10.1007/978-3-642-79451-3

[124] USEPA. Reference dose (RfD): Description and use in health risk assessments. Background document 1A. 1993. Available from: https://www.epa.gov/iris/reference-dose-rfd-description-and-use-health-risk-assessments [Accessed: November 28, 2023]

[125] Burton GA Jr. Sediment quality criteria in use around the world. Limnology. 2002;**3**:65-76. DOI: 10.1007/s102010200008

[126] Panagos P, Standardi G, Borrelli P, Lugato E, Montanarella L, Bosello F. Cost of agricultural productivity loss due to soil erosion in the European Union: From direct cost evaluation approaches to the use of macroeconomic models. Land Degradation & Development. 2018;*29*(3):471-484. DOI: 10.1002/ldr.2879

[127] Eisenhower DD. In: Eisenhower DD, editor. Military-Industrial Complex Speech. P Lillian Goldman Law Library. New Haven, CT: The Avalon Project; 1961. Available from: https://avalon.law.yale.edu/20th_century/eisenhower001.asp [Accessed: November 28, 2023]

[128] Medhurst MJ. Reconceptualizing rhetorical history: Eisenhower's farewell address. Quarterly Journal of Speech. 1994;**80**:195-218. DOI: 10.1080/00335639409384067

[129] Roosevelt FD. FDR's Fireside Chat on the Recovery Program. National Archives. Available from: https://www.archives.gov/education/lessons/fdr-fireside [Accessed: November 28, 2023]

Chapter 3

Beach Changes Triggered by Ground Subsidence and Waves Generated by Easterly Trade Wind near Pekalongan on North Shore of Java Island

Takaaki Uda, Susumu Onaka and Tomohiro Mori

Abstract

The shoreline recession caused by ground subsidence due to the excessive pumping of underground water and that caused by wind waves incident from the easterly direction under the tropical monsoon weather was investigated, taking the Pekalongan area on Java Island as a study area. The study area was divided into nine subareas on both sides of the Comal River delta, and the shoreline changes in each subarea were investigated using satellite images. Field observation was also carried out on August 10, 2022, in the Pekalongan area, and the height of ground subsidence was measured together with the observation of the inundated area using a UAV. It was found that, in this area, the height of ground subsidence reached approximately 1.6 m, and the sinking area has been rapidly expanding. In this area, predominant waves are constantly incident from the easterly direction under the tropical trade wind condition, and in some cases, waves are obliquely incident at a significantly large angle relative to the direction normal to the shoreline. Along the east coast of the Comal River mouth, the mean wave incidence angle was as large as 22° clockwise, whereas on the west side, the angle increased up to 42°, causing marked shoreline changes due to high-angle wave instability.

Keywords: ground subsidence, shoreline recession, Java Island, Pekalongan, high-angle wave instability, longshore sand transport

1. Introduction

A substantial proportion of the world's sandy coastline is already eroding, and such a situation may be exacerbated by climate change. Vousdoukas et al. [1] showed that these ambient trends in shoreline dynamics, combined with coastal recession driven by sea level rise, could result in the near extinction of almost half of the world's sandy beaches by the end of the century. Mentaschi et al. [2] gave a global and consistent evaluation of coastal morphodynamics between 1980 and 2015 based on satellite observations and concluded that the observed global trend in coastal erosion could be enhanced by future sea level rise along with the discussion on the impact due to anthropogenic factors. Thus, if a sea level rise develops, there will be an acceleration of existing beach erosion [3]. The changes predicted globally are already in evidence on sections of the world's coastline where land subsidence has produced a relative sea-level rise. Thus, the ground subsidence has the same effect as the sea level rise, implying that the study on ground subsidence is important to understand the impact of sea level rise to the coastal lowland. This could be done by a fact-finding investigation on real coasts.

Here, we aim to study the real example occurring between Pekalongan and Pemalang in Central Java, Indonesia, where severe ground subsidence has already occurred not only by the excessive pumping of underground water but also wave-induced sand transport under the constant tropical trade wind; topographic changes due to ground subsidence and wave action have simultaneously occurred. When a non-uniform ground subsidence occurs along the coastline, the longshore inclination of the shoreline changes, resulting in the occurrence of longshore sand transport, even though waves are incident normal to the previous shoreline. In this manner, the study on the combined effect of the ground subsidence and longshore sand transport to the topographic changes along a specific coast is no less important compared with the study on the global effect by sea level rise.

The coastline in the present study area between Pekalongan and Pemalang runs approximately in the east-west direction, and the shoreline protrudes into the Java Sea at the Comal River delta (**Figure 1**). Along this river delta coastline, waves may be incident at a large angle relative to the direction normal to the shoreline because of the protrusion of the shoreline, causing significant beach changes.

As regard to the ground subsidence in this area, Chaussard et al. [4] estimated the rate of ground subsidence in Pekalongan to be 3–10 cm/yr. in 2013. Then,

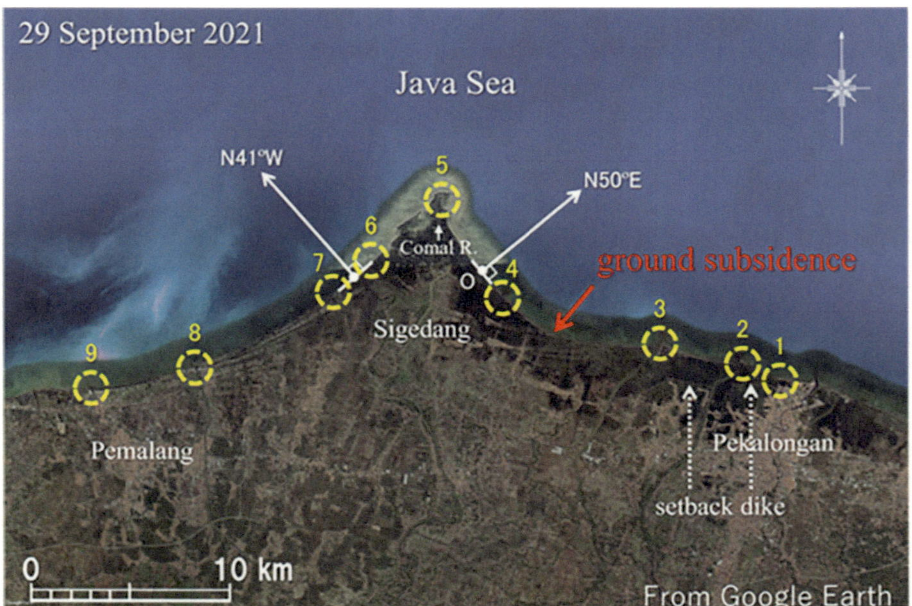

Figure 1.
Satellite image of the study area between Pemarang and Pekalongan, taken on 29 September 2021.

Andreas et al. [5] collected the evidence for subsidence based on repeated leveling measurements, GPS surveys, InSAR (Interferometric Synthetic Aperture Radar) measurements, extensometer, etc., and showed that the average rate of subsidence was as large as 0.01–0.1 m/yr. in 2018. They also pointed out that although dykes have been built against tidal inundation around the coasts on north Java coast, the dike subsiding through times results in the failure in their capability against ground subsidence and sea level rise, because the land subsidence and the sea level rise continue through times. Sarah and Soebowo [6] presented a brief review on land subsidence along the north coast on Java Island, and they showed the main features of the land subsidence including the subsurface geology, land subsidence monitoring, and analysis of possible causes and mechanism. Their time series analysis of InSAR images revealed that subsidence in Pekalongan occurred at a rate of 4.8–10.8 cm/year, suspected to be caused by groundwater withdrawal for agriculture.

Moreover, Andreas et al. [7] showed again in 2020 remote sensing capabilities on land subsidence and coastal water hazard using the GNSS (Global Navigation Satellite System) and InSAR data and concluded that, in the Pekalongan area, the ground subsidence started since late 1970s and the entire ground subsidence reached 3.7 m until 2015.

Sidiq et al. [8] also investigated the ground subsidence of the entire Java Island using SAR satellite images and argued that the height of ground subsidence in Pekalongan reached 22 cm between 2017 and mid-2020. Thus, most of the studies focus on the rate of ground subsidence in the inland area of Pekalongan, and studies on the interaction between cumulative ground subsidence and beach changes caused by the imbalance in longshore sand transport due to waves are rare.

Similar ground subsidence triggered by the excessive pumping of underground water has occurred in the southern part of Kujukuri coastal plain in Chiba Prefecture, Japan, and the maximum ground subsidence reached 90 cm between 1968 and 2012, causing the shoreline recession of 18 m when assuming the foreshore slope of 1/20 [9]. When the effect of sea level rise associated with the greenhouse effect occurs with the ground subsidence, enormous damage may occur. Therefore, in this study, beach changes were investigated, taking the suburbs of Pekalongan as an example, where coastal lowland vulnerable for coastal damage develops along the coastline, and both the ground subsidence and the shoreline recession caused by the imbalance in longshore sand transport due to waves simultaneously have occurred. Through the study, we investigate not only the expansion of the inundation area associated with ground subsidence but also the reversal in the direction of longshore sand transport caused by local ground subsidence. Moreover, along the study area, the angle of the predominant waves takes a value greater than 45°, possibly causing high-angle wave instability. We also discuss a measure to prevent the devastation of the coastline along such a coast.

First, nine subareas were adopted alongshore between Pekalongan and Pemarang located 20 km west of the Comal River delta (**Figure 1**), and the shoreline changes in each subarea were investigated using Google Earth, together with field observation on 10 August 2022 near Pekalongan. In **Figure 1**, the lowland shown in black extends in the vicinity of the shoreline in subareas 1–4 west of Pekalongan where significant ground subsidence occurred and particularly large ground subsidence occurred in subareas 2 and 4.

The results of the study in each subarea are described in Chapter 3. Then, in Chapter 4, the longshore distribution of the angle relative to the direction normal to the shoreline is shown, which closely relates to the longshore sand transport on each coast. Then, discussions were made in Chapter 5.

Regarding the wave characteristics in this area, wave height was measured by Sakuno et al. [10] using the satellite sea level altimeter. They showed that the significant wave height with a high probability of occurrence in the Java Sea ranges between 0.5 and 1.0 m, relatively low with a wave period shorter than 4 s. Furthermore, the significant wave height H_{95} with the non-exceedance probability of 95% was 1.48 m.

2. Method of study

The Comal River delta protrudes into the Java Sea in the central part of the study area (**Figure 1**). In this area, subareas 1–4 were set westward from Pekalongan east of the Comal River delta, whereas subareas 6–9 were set alongshore until Pemalang west of the river delta along with subarea 5 at the center of the Komal River delta. The shoreline of the Comal River delta has an asymmetric shape with respect to the centerline of the delta, as shown in **Figure 1**. A straight line is drawn in the east–west direction at a location 4.2 km south of the river mouth, and the direction normal to the shoreline at the intersections between this line and the shoreline is calculated; the angle of the direction normal to the shoreline with respect to the north becomes 50° (N50°E) on the east side and 41° (N41°W) on the west side of the delta. Thus, the angle of the direction normal to the shoreline relative to the north on the east side (50°) is larger than that of 41° on the west side of the river delta. This asymmetry of the river delta is assumed to be due to the asymmetric effect of waves incident from the eastern side of the river delta under the constant action of easterly wind. Because when waves are incident from the direction of the center line dividing a symmetric river delta into two, a symmetric river delta must be formed due to the symmetric action of waves.

In this study, the suburbs of Pekalongan (subarea 2) where severe ground subsidence occurred, the Sikebu River mouth (subarea 4), the Comal River delta (subarea 5), and subarea 6 west of the river delta, as shown in **Figure 1**, were selected out of nine subareas, and the shoreline changes in each subarea were investigated. The distribution of wave incidence angle relative to the direction normal to the shoreline was measured in these nine subareas because oblique wave incidence is vital to cause the longshore sand transport.

3. Results of the study in each subarea

3.1 Subarea 1 (suburbs of Pekalongan)

Figure 2 shows a satellite image of the area immediately west of Pekalongan taken on 29 September 2021. The residential area of Pekalongan extends along the left bank of the Banger River. Severe ground subsidence occurred in this area, and the urban area is surrounded by the submerged area shown in black. In addition, jetties of approximately 300 m length have been constructed on both sides of the Banger River, and the shoreline advances most on the east side of the jetty, implying that westward longshore sand transport prevails in this area.

On the other hand, the shoreline was assumed to retreat at the west end of the seawall because the shoreline in the urban area west of the river mouth is fixed by the seawall. Therefore, when adopting a rectangular area in **Figure 2** west of the urban

Figure 2.
Satellite image of the area near Pekalongan and detailed study area.

area, satellite images taken five times between March 2003 and September 2021 were compared (**Figure 3**).

In March 2003, no clear evidence of erosion or effects of ground subsidence were observed, and the shoreline extended straight in the east-west direction, and a sandy beach extended alongshore with neatly divided gardens and fish ponds in the hinterland (**Figure 3(a)**). However, by September 2009, groins and a seawall were constructed, as shown in **Figure 3(b)**, east of the crematory indicated by a circle in **Figure 3(a)** as a measure against erosion. Then, because of the ground subsidence and the blockage of westward longshore sand transport, the shoreline was retreated by 54 m at the west end of the seawall by August 2013 (**Figure 3(c)**). Here, in **Figure 3(c)**, the propagation of swell waves from the northeast can be clearly identified; therefore, the direction normal to the wave crests immediately offshore of the wave breaker zone near the crematory was determined to be N36°E. Because the direction normal to the shoreline in this vicinity is N14°E, waves are obliquely incident clockwise at an angle of 22° relative to the direction normal to the mean shoreline, resulting in the predominance of northwestward longshore sand transport and erosion at the west end of the seawall. Moreover, ground subsidence was intensified between 2009 and 2013, and footpaths of the fish pond became invisible.

Then, a straight seawall was constructed at a location 30 m offshore of the shoreline west of the crematory as a measure against erosion until April 2014 and connected to the seawall constructed earlier (**Figure 3(d)**). However, these seawalls lost their function because of ground subsidence, and they were destroyed by September 2021 (**Figure 3(e)**). Simultaneously, severe ground subsidence occurred, and farmlands and fish ponds that extended in the coastal lowland in 2003 were submerged under the sea surface and the facilities of the crematory remained under the sea.

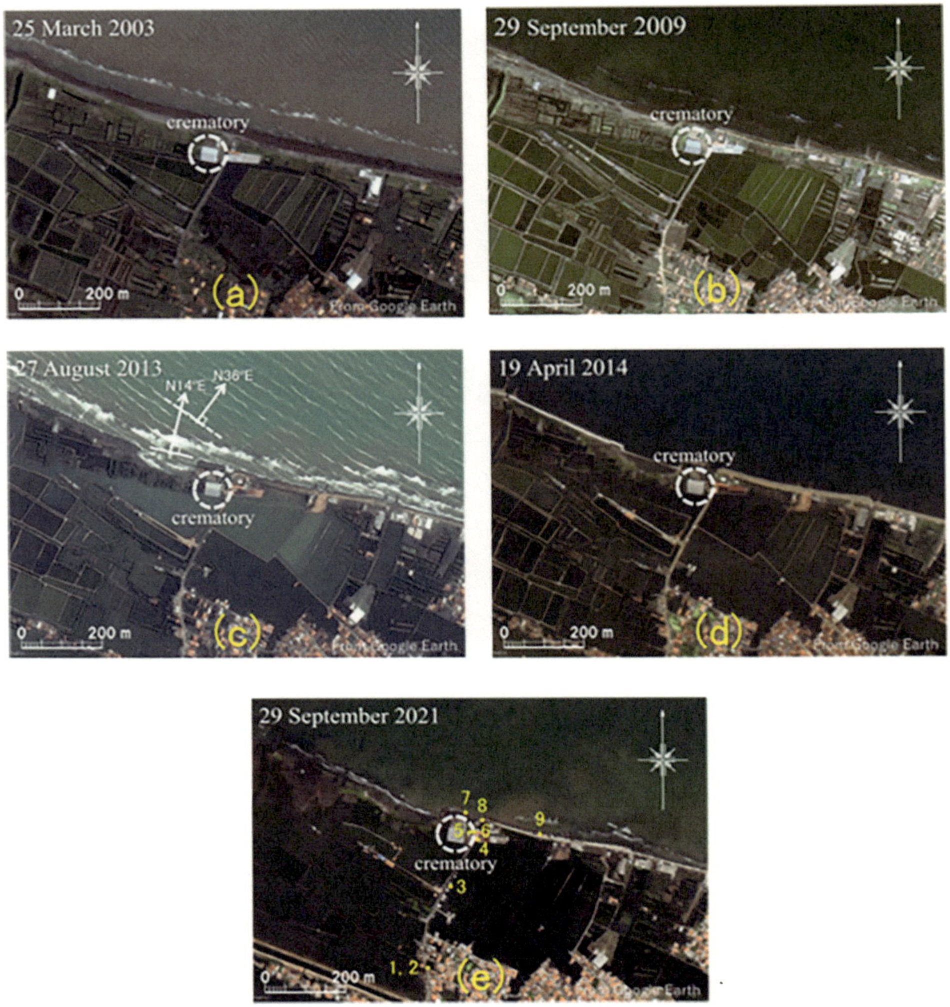

Figure 3.
Satellite images of subarea 1 at Pekalongan.

As mentioned above, marked topographic changes occurred between March 2003 and September 2021 in subarea 1. The situation of the ground subsidence was confirmed on August 10, 2022, by the field observation. **Figure 3(e)** shows the location (Sts. 1–9) where photographs were taken. First, at St. 1 shown in **Figure 3(e)**, the difference in elevation between the sea surface at the subsided land on the right (west) side of the seawall and the surface of the coastal road on the left was measured using a measuring stick (**Figure 4(a)**). As a result, the elevation of the road surface was found to be 0.2 m lower than that of the sea surface at that time. As the tide level at the measurement time was the mean sea level (MSL) + 0.10 m, the ground elevation of the road was MSL − 0.10 m.

Similarly, **Figure 4(b)** shows the measurement of the ground elevation on the east side of the road, and the ground elevation was found to be 0.8 m lower than the elevation of the road. Although the elevation of the road has been raised after the successive ground subsidence, the ground elevation of the residential area has not been

Figure 4.
Photographs showing ground subsidence taken at Sts.1–9 in Pekalongan.

changed; these indicate that the ground elevation in this vicinity is MSL − 0.90 m. As the ground elevation before the ground subsidence was not known, the accurate ground subsidence could not be determined. However, when the original ground elevation of the coastal lowland was assumed to be equal to the berm height of MSL + 0.7 m, described later, the ground subsidence was determined to be 1.6 m.

Figure 4(c) taken at St. 3 near the crematory further seaward of St. 2 shows the condition of the seawall. The seawater spouted out through a small hole of the seawall to the road on the east (right) side of the seawall of 0.8 m height. It is clearly understood from this photograph that the sea level was higher than the road level. Furthermore, **Figure 4(d)** taken at St. 4 shows the east side of the crematory, after walking straight along the road toward the sea. From this photograph, it seems that the crematory is floating on the sea with a mangrove forest.

Then, the situation in the vicinity of the crematory was observed in detail, which showed that almost all the land on the east side of the crematory subsided below the sea surface, and the trees were submerged below the sea surface, as shown in **Figure 4(e)**. This crematory was located approximately 50 m landward of the shoreline in 2003, as shown in **Figure 3(a)**, but the crematory sunk below the sea surface because of the ground subsidence and the sand discharge due to the westward longshore transport.

However, a low berm was formed near the retreated shoreline. Therefore, the berm height was measured relative to the sea level at 12:35 on 10 August 2022, which showed that the berm height was 0.6 m (**Figure 4(f)**). As the elevation of the sea surface at the measurement time was MSL + 0.10 m, the berm height relative to the

Figure 5.
Oblique image of the study area taken on 1 July 2022.

MSL was 0.7 m. In addition, the vertical seawall protecting the seaward side of the crematory was severely damaged by wave action along with the inundation behind the seawall owing to wave overtopping (**Figure 4(g)**).

When the condition of a road extending along the east side of the crematory, as shown in **Figure 3(e)**, was investigated after the observation near the crematory, the road was found to be protected against waves and erosion by both a 0.8-m-high vertical seawall and the zig-zag-shaped revetment in front of the seawall (**Figure 4(h)**). However, the seawall was severely destroyed by waves with no trace of its original shape remaining (**Figure 4(i)**).

Figure 5 shows an oblique image of the measurement carried out on July 1, 2022, using a UAV (drone: DJI Phantom 4 Pro V2.0). This image clearly shows that ground subsidence occurred in an extensive area along the road, and the ground on both sides of the road subsided below the sea surface. Although a slender mangrove forest remained west of the crematory, all the surrounding areas were submerged below the sea surface, and the seawall constructed along the coastline was invisible.

3.2 Subarea 4 (Sikebu River mouth)

The Sikebu River in subarea 4 flows into the Java Sea at a location 8.6 km west of the Semut River in subarea 3. The coastline inclination near this river mouth forms an angle of 37° clockwise relative to that of the mean coastline between subareas 1 and 3, as shown in **Figure 1**. Therefore, under the condition that waves are incident from the northeast, the angle between the direction normal to the shoreline and the wave direction becomes smaller than those in subareas 1–3, implying the decrease in longshore sand transport compared with those in subareas 1–3.

The satellite image taken in February 2005 is shown in **Figure 6(a)**. At that time, the Sikebu River significantly meandered and then flowed out to the Java Sea near the river mouth because the mouth was closed by a sand spit extending southeastward. Here, wave crestlines can be clearly observed offshore of the shoreline in this satellite image. Therefore, the wave incident angle was determined to be N27°E from the direction normal to the crestline offshore of the shoreline. As the direction normal to the mean shoreline at that time was N50°E, as shown in **Figure 6(a)**, waves were incident at an angle of 23° counterclockwise relative to the direction normal to the shoreline in this area, causing southeastward longshore sand transport, which is different from the predominance of westward longshore sand transport in subareas 1–3. The reversal in the direction of longshore sand transport from westward in subareas 1–3 to southwestward in subarea 4 is due to the change in the shoreline angle; the

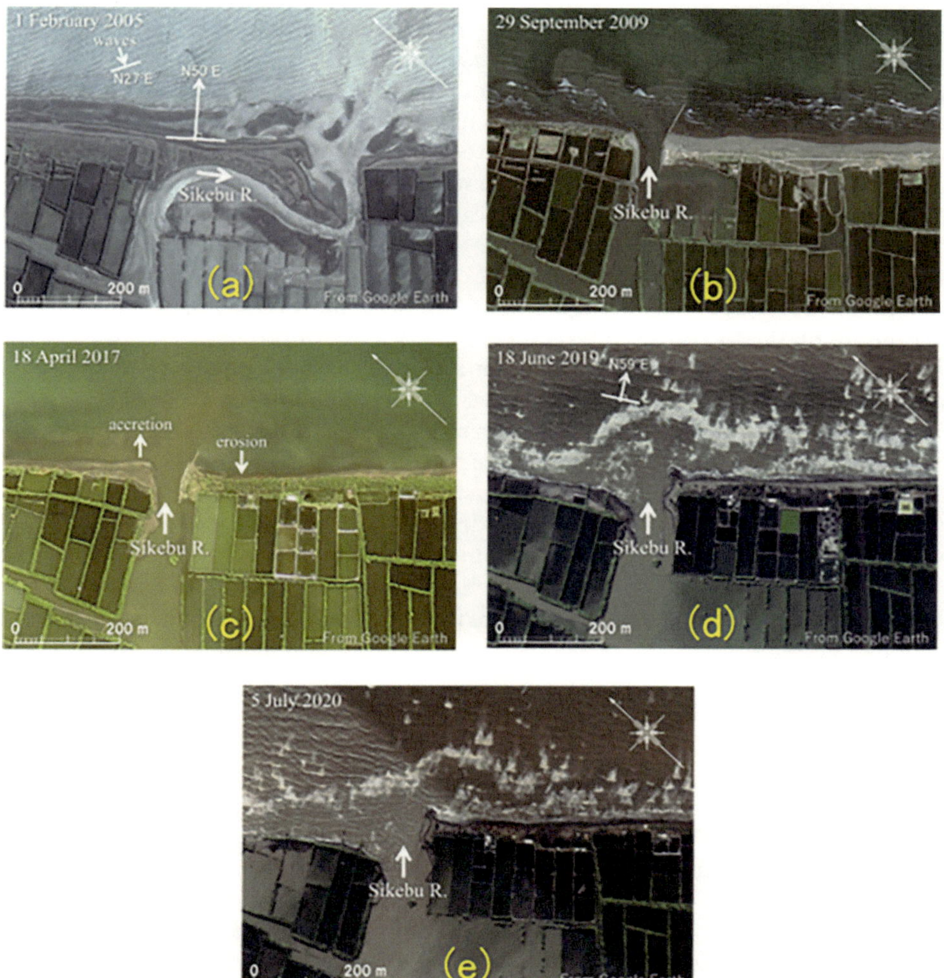

Figure 6.
Satellite images of subarea 4 (Sikebu River mouth).

coastline near subarea 4 forms an angle of 37° clockwise relative to that of the mean coastline between subareas 1–3 because subarea 4 is located on the eastern side of the Comal River delta significantly protruding northward, and waves from the eastern side are incident counterclockwise relative to the direction normal to the shoreline.

Figure 6(b) shows a satellite image taken in September 2009. The river, which meandered significantly southeastward, became straight because of the extension of two jetties at the river mouth. The south jetty was asymmetrically arranged with the south jetty being longer than the north jetty by approximately 70 m. Therefore, the shoreline may locally advance on the southeast side of the jetty owing to the wave diffraction effect of the jetty, but the shoreline formed was straight.

Then, until April 2017, the shoreline retreated in the area south of the jetty, whereas on the north side of the river mouth, sand was deposited, resulting in the shoreline reaching the tip of the north jetty (**Figure 6(c)**, although the south jetty was shortened relative to that in 2009 (**Figure 6(b)**). The fact that the shoreline advanced on the north side of the jetty and retreated on the south side clearly indicates that southeastward longshore sand transport prevailed at this river mouth until 2017.

However, the shoreline on the north side of the river mouth was retreated by 80 m by June 2019, although the shoreline on the south side did not change (**Figure 7(d)**). Here, wave crestlines can be observed offshore of the left bank of the river mouth. Thus, the wave incident angle was determined from the satellite image; the angle of N59°E was determined from the direction normal to the crestline offshore of the shoreline, as shown in **Figure 6(d)**. Thus, the wave angle was found to rotate clockwise by 32° relative to the initial wave angle (N27°E) in 2005 because of severe ground subsidence.

Near the river mouth, many small roads separating fishponds disappeared owing to the ground subsidence. The same condition can be seen from the satellite image taken in July 2020, as shown in **Figure 6(e)**. The shoreline on the north side of the river mouth markedly retreated because of significant ground subsidence west of the Sikebu River. When comparing the shoreline in April 2017 (**Figure 6(c)**) with that in July 2020 (**Figure 6(e)**), the shoreline recession on the north side of the river mouth reached 110 m until July 2021 because of marked ground subsidence. Thus, marked topographic changes occurred owing to the ground subsidence since 2019 at the Sikebu River mouth, causing the reversal in the direction of longshore sand transport, although southeastward longshore sand transport was predominant in 2005. It is concluded that when the non-uniform ground subsidence occurs along the coastline, the longshore inclination of the shoreline changes, resulting in the occurrence of longshore sand transport, even if waves are incident from the direction normal to the previous shoreline.

3.3 Subarea 5 (Comal River)

The shoreline changes in subarea 5 around the Comal River mouth were similarly investigated using the satellite images taken five times between September 2009 and May 2022 (**Figure 7**). The images show that this river flowed northward in September 2009 with an opening at the north end. It is found that the sediment supplied from the river was deposited around the river mouth, forming a shallow offshore terrace (**Figure 7(a)**). On the other hand, the amount of sand deposited on the west side of the river was smaller than that on the east of the river mouth, and an extensive lagoon enclosed by a slender barrier island was formed. The greater deposition of sand on the east side indicates that the sediment supplied from the Comal River was mainly transported to the east coast.

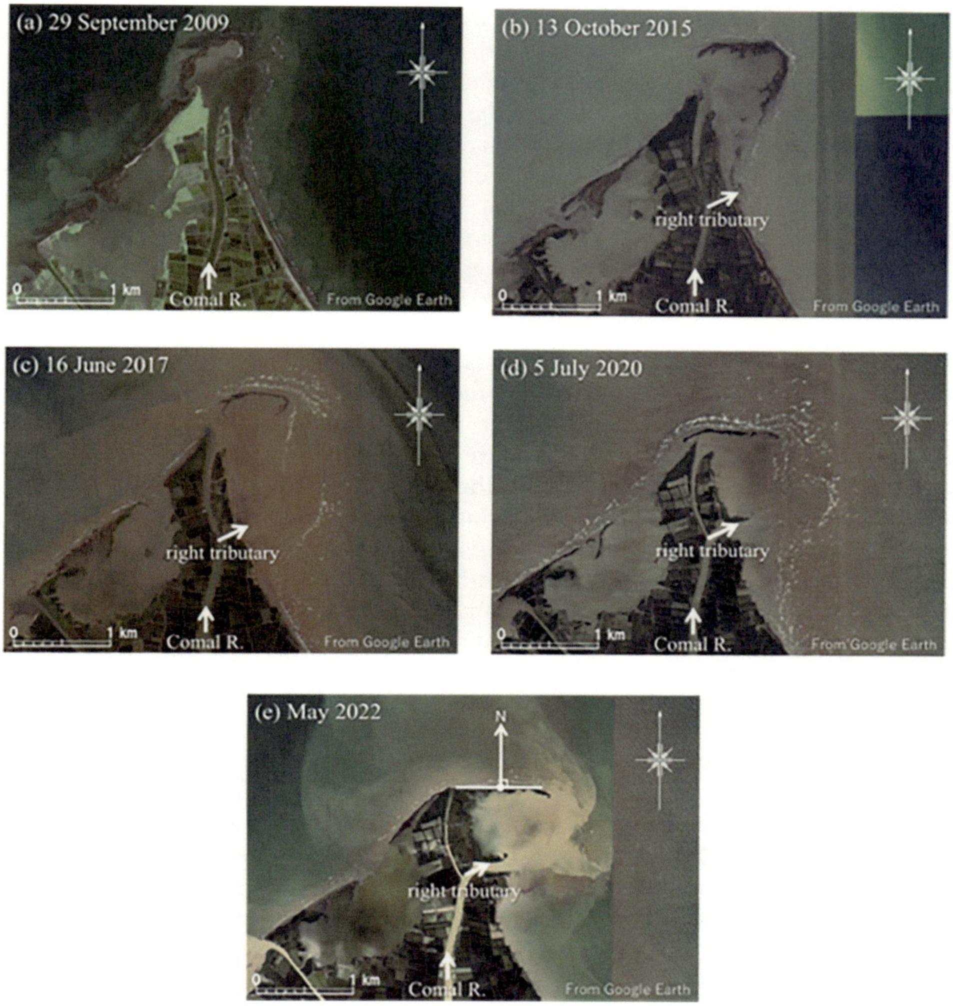

Figure 7.
Satellite images of subarea 5 (Comal River mouth).

Then, in 2015, the right tributary was formed at a location 1.1 km south of the river mouth of the main stream (**Figure 7(b)**). At this time, an umbrella-shaped sandbar was remained offshore of the mouth of the main stream, and two slender sandbars have extended along the marginal line of the offshore terrace: one is westward and the other is southwestward. The effect of the main stream was assumed to be weakened because of the decrease in discharge, and the offshore terrace started to deform mainly owing to wave action.

In June 2017, although a shoal still remained offshore of the main stream, the size of the sandbar extending in the east–west direction offshore of the river mouth decreased, and the sandbar moved southward by approximately 120 m (**Figure 7(c)**). As the sediment was also supplied from the right tributary, another shoal started to be formed offshore of the mouth of this tributary. The formation of this shoal can be recognized from the formation of white-capped breaker lines along the external line of the shoal.

In July 2020, the slender sandbar located offshore of the mouth of the main stream moved by 400 m southward compared with its previous location in October 2015 because the amount of discharge of the right tributary increased (**Figure 7(d)**). At the same time, the increase in the size of the terrace offshore of the mouth of the right tributary can be seen from the distribution of the breaker zone. Then, until May 2022, the sandbar extending in the east–west direction offshore of the main stream further moved southward by approximately 180 m while maintaining its slender shape, and the river mouth of the main stream was completely buried with sand. Thus, since 2015, when the effect of the mainstream was weakened, a slender sandbar extending in the east–west direction offshore of the main stream has moved southward while maintaining its direction.

For the parallel movement of a slender sandbar while maintaining its shape, it is necessary that the predominant waves are incident from the direction normal to the shoreline of the slender sandbar. Therefore, the direction normal to the shoreline was determined from the satellite image (**Figure 7(e)**), which was found to be northward approximately parallel to the flow axis of the Comal River. Thus, the Comal River flowed northward until 2009 with the sediment being supplied north of the river mouth. However, because of the excavation of the right tributary by May 2022, all the discharge of the river flowed through this tributary. As a result, the mouth of the main stream was completely closed by a sandbar, and an offshore slender sandbar connected to the right bank of the river mouth was formed. The slender sandbar offshore of the river mouth moved southward owing to the action of waves incident from the north, while maintaining its shape.

3.4 Subarea 6 (Kauman River mouth)

The Kauman River mouth is located 5.2 km southwest of the main stream of the Comal River. In this area, the westward extension of the river mouth bar was investigated from the satellite images taken three times between June 2014 and July 2020 (**Figure 8**). In June 2014, the sandbar of the right bank extended by 69 m from point A at the west end of the mangrove forest (**Figure 8(a)**). Until August 2017, the sandbar elongated westward by 250 m west of A. During this period, the smooth sand supply from the right to the left bank was assumed to be interrupted owing to the extension of the sandbar on the right bank.

When a sand spit is formed at the tip of the barrier island, the sand spit itself develops owing to the continuous supply of sand from the upcoast, and almost all sand is used for the elongation of the sand spit, resulting in the disruption of continuous sand movement along the shoreline because of the discontinuity in shoreline at the tip. Noshi et al. [11] showed an example of rapid development of a recurved spit at the south end of Phan Rang Bay located in the southeast part of Vietnam, and the beach was eroded on the opposite side of the tip of the sand spit during the elongation of the sand spit. The same examples were observed along the curved shoreline north of Pengambengan fishing port located in the western part of the Bali Strait [12] and along the barrier island developed in Nakatsu tidal flat, Japan [13]. Accordingly, the same phenomenon is assumed to occur at the southeast end of the barrier island elongating along the Kauman River mouth.

Furthermore, in **Figure 8(b)**, another sandbar started to elongate westward while covering the previous sandbar. This sand spit further elongated up to a location 520 m west of point A by July 2020 (**Figure 8(c)**). Here, in **Figure 8(c)**, many wave crest lines can be seen offshore of the breaker zone; therefore, the wave angle can

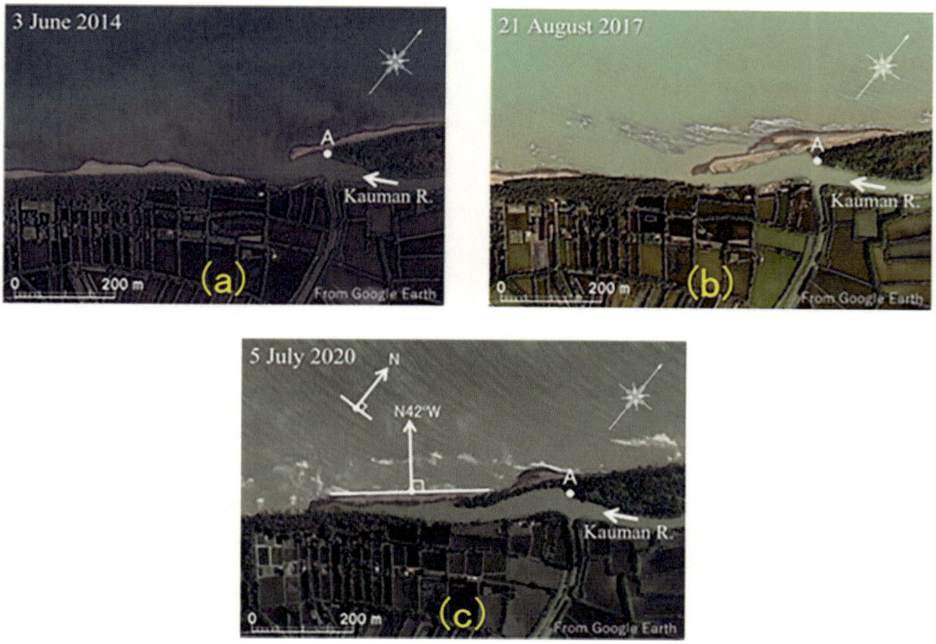

Figure 8.
Satellite images of subarea 6 (Kauman River mouth).

be determined from the shape of these crestlines; the wave direction indicated the north. On the other hand, when a line normal to the shoreline was drawn at the center of the long sand spit, the direction of this normal line became N42°W; waves were obliquely incident clockwise by 42° relative to the direction normal to the shoreline, and this condition is close to the angle of 45° at which the shoreline instability will occur owing to high-angle wave instability [14]. For this reason, intensive westward longshore current (sediment transport) was induced in subarea 6, resulting in the elongation of a sand spit.

3.5 Subarea No. 9 (Wirasa River)

The Wirasa River flows into the Java Sea at a location 3.8 km northwest of Pemalang. **Figure 9(a)** shows the satellite image of this river taken in October 2004. The parallel jetties were already constructed at the river mouth by October 2004, and the east jetty was longer than the west jetty. As the east jetty blocked longshore sand transport, the shoreline on the east side of the river advanced offshore relative to that on the west side, indicating the predominance of westward longshore sand transport. When drawing a line normal to the shoreline at a location 130 m east of the east jetty, the direction became N11°E. Similarly, the direction normal to the shoreline at a location 370 m east of the east jetty that was not affected by the blockage of longshore sand transport can be determined to be north. Finally, it is concluded that waves are incident at an angle of 11° clockwise at this river mouth. As the wave diffraction zone is formed west of the jetties against waves incident from the N11°E direction, a bay-shaped shoreline was formed immediately west of the west jetty.

Similarly, **Figure 9(b)** shows the satellite image taken in September 2013. Because the east jetty was extended up to a location 220 m offshore of the seawall, the wave

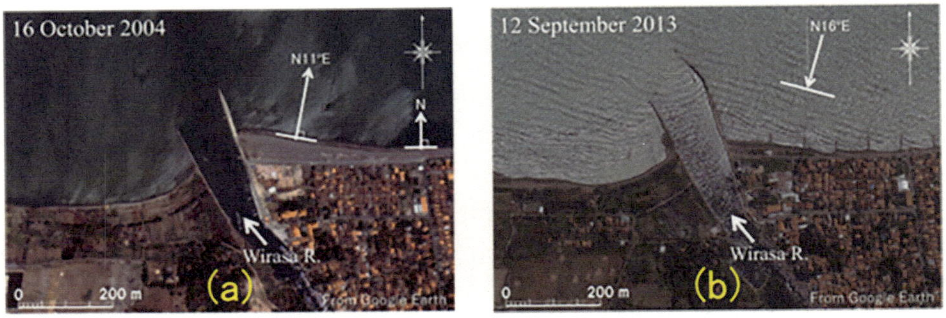

Figure 9.
Satellite images of subarea 9 (Wirasa River mouth).

diffraction effect was intensified, causing the shoreline to advance by approximately 40 m west of the jetty. Here, wave crest lines can be seen at a location 130 m offshore of the east jetty; therefore, the wave incidence angle can be measured to be N16°E. This angle is close to that of N11°E obtained in **Figure 9(a)**.

4. Longshore sand transport and shoreline instability

The incident wave angle in subareas 1–9 set along the shoreline of the Comal River delta was evaluated by two methods. First, the wave direction was determined from the direction normal to the wave crestline offshore of the breaker zone shown in satellite images (Type I). Second, the wave direction was determined from the difference in angle between the directions normal to the shoreline before and after the construction of a groin (Type II). **Table 1** shows the wave angle determined in each subarea by Type I or II method.

Table 1 indicates that the mean wave angle in subareas 1–3 east of the Comal River delta was as large as 22° clockwise on average; therefore, marked westward longshore sand transport could occur. In contrast, in subarea 4, immediately east of the Comal

Number of subarea	Wave direction	Wave angle	Wave angle	Date of taking satellite image
1	N36°E	22°	I	27 August 2013
2	N34°E	20°	II	19 April 2014 and 29 September 2021
3	N40°E	23°	II	4 May 2020
4	N27°E	−23°	I	1 February 2005
5	N	0°	II	May 2022
6	N	42°	I	5 July 2020
7	N	35°	I	5 July 2020
8	N12°E	22°	II	16 June 2017
9	N16°E	16°	I	16 October 2004

Table 1.
Wave angles at subareas 1–9.

DOI: http://dx.doi.org/10.5772/intechopen.1004907

Wave conditions	**Incident waves: $H_I = 2$ m, $T = 8$ s, wave direction $\theta_I = 55°$ relative to x-axis**
Berm height	$h_R = 2$ m
Depth of closure	$h_c = 6$ m
Equilibrium slope	$\tan\beta_c = 1/10$
Coefficients of sand transport	Coefficient of longshore sand transport $K_s = 0.01$ Coefficient of Ozasa and Brampton term $K_2 = 1.62K_s$ Coefficient of cross-shore sand transport $K_n = K_s$
Mesh size	$\Delta x = \Delta y = 10$ m
Time intervals	$\Delta t = 0.8$ h
Duration of calculation	3.2×10^4 h (4×10^4 steps)
Boundary conditions	Shoreward and landward ends: $q_x = 0$ Right and left boundaries: $dq_y/dy = 0$ Lower right corner of beach edge: $dq_x/dx = 0$ or $dq_y/dy = 0$

Table 2.
Calculation conditions.

River mouth, the wave angle was −23° in 2005 before the occurrence of the marked ground subsidence, which was the reversed direction relative to those in subareas 1–3. However, after the ground subsidence, the wave angle rotated by 30° in this area, causing the reversal in the direction of longshore sand transport. In addition, in subarea 5 at the tip of the river delta, the wave direction was from the north.

In subarea 6 west of the river delta, the wave angle was as large as 42°, which is sufficiently large to cause the development of shoreline undulation due to the mechanism of high-angle wave instability [14]. In domains, such as subarea 6, when a groin is extended to block longshore sand transport, a stable shoreline cannot be formed upcoast of the groin, and longshore sand transport may flow out downcoast turning around the tip of the groin [15, 16].

Uda and Serizawa [16] carried out a numerical simulation of the extension of a sand spit around groins constructed on a coast satisfying the condition of high-angle wave instability using the BG model [17], taking the coast of Narvacan in Lingayen Gulf in the Philippines. A model beach with a uniform profile ranging in the depth zone between the berm height (h_R) and the depth of the closure (h_c) was set up on the upstream end on a solid bed, as shown in **Figure 10**, with sand being supplied from the upcoast boundary between h_R and h_c, and subsequent beach changes due to wave action were predicted. A solid flat bed with the depth of h_c was assumed in the offshore zone. A seawall was set along the coastline concurrently with the installation of eight groins of 490 m length at 90 m intervals. Incident wave height was assumed to be 2 m, which is equal to the berm height, and wave period was assumed to be 8 s. Waves were assumed to be incident from the clockwise direction of $\theta = 55°$. The wave field was calculated using the energy balance equation. The calculation was carried out up to 3×10^4 steps, where 10^4 steps correspond to 0.9 years. The other calculation conditions are shown in **Table 2**.

Figure 10 shows the results of the calculation. Because waves are obliquely incident from the clockwise direction at an angle of 55°, leftward longshore sand transport developed, and sand supplied from the northwest (right) boundary was deposited while forming a sand body, which moved southeastward over the tips of the groins. There is a gap between the tip of the groin and the downcoast shoreline at

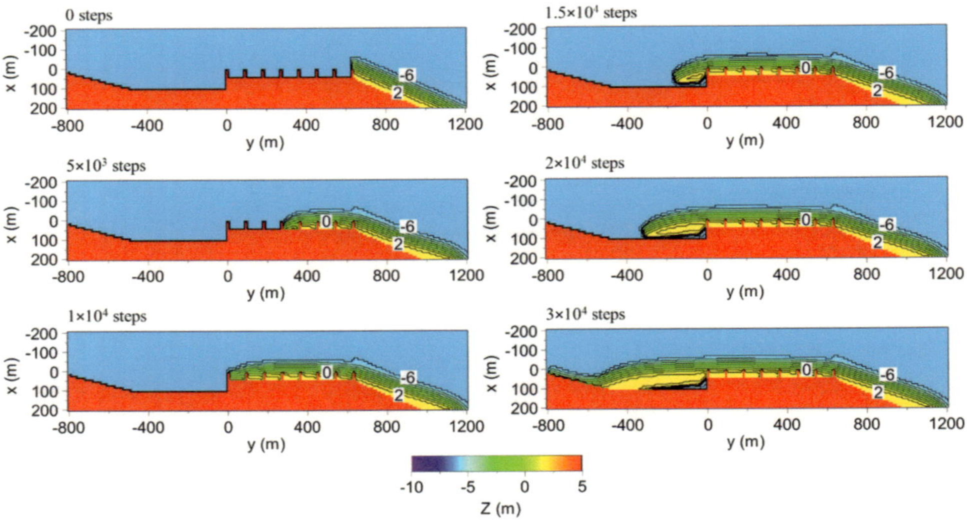

Figure 10.
Prediction of development of a sand spit around the tip of groins when waves are obliquely incident to the direction normal to the shoreline at an angle of 55° [16].

the groin located at the southeast end, so a slender sand spit elongated southeastward (leftward) while leaving a lagoon behind it. Then, a barrier island was formed after the elongation of the slender sand spit. These results corresponded well with the measured changes [16]. Thus, it is concluded that a statically stable shoreline cannot be formed using groins under high-angle wave instability, because longshore sand transport may flow out downcoast turning around the tip of the groin. It is required to maintain continuous sand movement on such a coast satisfying high-angle wave instability.

In areas further west, the wave angle in subarea 7 located 7.7 km southwest of the Comal River mouth is 35°, and those in subareas 8 and 9 are 22 and 16°, respectively. Although the oblique wave angle decreases with distance from the Comal River mouth, waves are obliquely incident clockwise at a large angle at any points, causing westward longshore sand transport. Therefore, when a jetty is extended on such a coast, careful attention on downcoast erosion is required.

5. Discussion

Along the shoreline of 33 km length between Pekalongan and Pemalang in Central Java, nine subareas were adopted from both sides of the Comal River delta, and the characteristics of longshore sand transport in each subarea were investigated using satellite images. As a result, it was found that the wave angle relative to the direction normal to the shoreline is large at 22° on average in subareas 1–3 east of the Comal River delta, causing significant beach changes due to westward longshore sand transport. In subarea 4 (Sikebu River mouth), the reversal of longshore sand transport was observed because of the marked ground subsidence.

Finally, the mechanism is schematically shown in **Figure 11**. When the wave angle is sufficiently small, the shoreline between the groins is stabilized for the shoreline to be normal to the wave direction (**Figure 11(a)**). However, when high-angle wave instability occurs, the volume of sand to be deposited upcoast of groin G_1 is minimal, and almost all the sand discharges downcoast and the shoreline retreats up to the seawall (**Figure 11(b)**), resulting in the formation of flying spits [14, 16].

For a soft measure against erosion to be adopted on such a coast, the continuity of longshore sand transport must be maintained (**Figure 11(c)**). Thus, in subarea 6, the only way to maintain the present shoreline is to form a dynamically stable shoreline by maintaining the continuity of longshore sand transport, that is, sand bypassing.

Apart from beach changes caused by longshore sand transport, severe ground subsidence due to the excessive pumping of ground water occurred in subareas 1–4 west of Pekalongan with a maximum subsidence of the ground reaching approximately 1.6 m in height, and the submerged coastal land had been expanding. For the measures against the submergence of the ground and erosion by the spatial imbalance in longshore sand transport, the setting back of the seawall is under way at present together with the installation of pump stations to discharge seawater for the reduction in water level in subareas 1–4. Accordingly, not only short-period measures should be taken but also the pumping of the underground water, which is thought to be the primary cause of the ground subsidence, should be controlled in these areas. In addition to these, mangroves should be planted to form a forest in an extremely shallow sea offshore of the seawall to enhance wave attenuation. In the areas west of the Comal River mouth, the effect of ground subsidence was not very marked, but waves

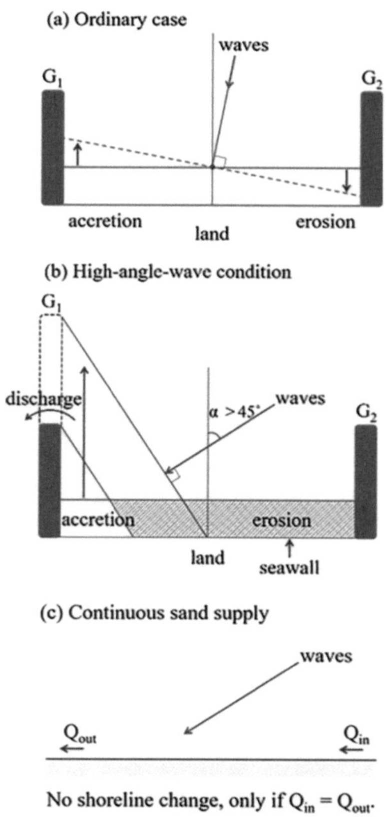

Figure 11.
Schematic of shoreline changes around groins.

were obliquely incident at a large angle from the eastern side. Therefore, maintaining the continuity of longshore sand transport is required in the long-term methods such as sand bypassing.

6. Conclusions

It was found that, in the Pekalongan area, the height of ground subsidence reached approximately 1.6 m which is much greater than the expected sea level rise of 0.6 m in the coming 100 years reported in IPCC [18], and the sinking area has been rapidly expanding at present. Together with the ground subsidence, the predominant waves constantly developing under the easterly trade wind are incident to the coast, causing a significant shoreline recession downcoast of the structures. Along the coastline east of the Comal River mouth, the mean wave incidence angle was as large as 22° clockwise, whereas on the west side, the angle increased up to 42°, causing marked shoreline changes due to high-angle wave instability, resulting in significant beach changes. Therefore, in this area, maintaining the continuity of longshore sand transport is required in the long-term methods such as sand bypassing instead of structural defense using groins or detached breakwaters.

Acknowledgements

This study is part of a JICA study related to the Project for Planning Coastal Master Plan Against Beach Erosion on North Coasts of Java. We would like to thank JICA for the permission to present the results.

Author details

Takaaki Uda[1*], Susumu Onaka[2] and Tomohiro Mori[2]

1 Public Works Research Center, Nihon University, Tokyo, Japan

2 Nippon-Koei Co., Ltd., Tokyo, Japan

*Address all correspondence to: uda@pwrc.or.jp

References

[1] Vousdoukas M, Ranasinghe R, Mentaschi L, Plomaritis TA, Athanasiou P, Luijendijk A, et al. Sandy coastlines under threat of erosion. Nature Climate Change. 2020;**10**:260-263

[2] Mentaschi L, Vousdoukas MI, Pekel J-F, Voukouvalas E, Feyen L. Global long-term observations of coastal erosion and accretion. Scientific Reports. 2018;**8**:12876

[3] Bird ECF. Coastal erosion and rising sea-level, 87-103, chapter 4. In: Milliman ID, Haq BU, editors. Sea-Level Rise and Coastal Subsidence. London: Kluwer Academic Publishers; 1996

[4] Chaussard, E., Amelung, F., Abidin, H. and Hong, S. H. 2013. Sinking cities in Indonesia: ALOS PALSAR detects rapid subsidence due to groundwater and gas extraction, Remote Sensing of Environment, 128, 150-161, 2013.

[5] Andreas H, Abidina HZ, Sarsito DA, Pradipra D. Insight analysis on dyke protection against land subsidence and the sea level rise around northern coast of Java (PANTURA) Indonesia. Geoplanning: Journal of Geomatics and Planning. 2018;**5**(1):101-114. DOI: 10.14710/geoplanning

[6] Sarah D, Soebowo E. Land subsidence threats and its management in the north coast of Java. IOP Conference Series: Earth and Environmental Science. 2018;**118**:1-8. DOI: 10.1088/1755-1315/118/1/012042

[7] Andreas H, Abidin HZ, Sarsito DA, D. A., Pradipta, D. Remote sensing capabilities on land subsidence and coastal water hazard and disaster studies. IOP Conference Series: Earth and Environmental Science. 2020;**500**:1-12. DOI: 10.1088/1755-1315/500/1/012036

[8] Sidiq TP, Gumilar I, Meilano I, Abidin HZ, Andreas H, Permana A. Land subsidence of Java north coast observed by SAR interferometry. IOP Conference Series: Earth and Environmental Science. 2021;**873**:1-8. DOI: 10.1088/1755-1315/873/1/012078

[9] Uda T, Yoshida R, Todoroki T. Beach Changes Triggered by Imbalance of Longshore Sand Transport and Ground Subsidence on South Kujukuri Beach. World Scientific Publishing, Coastal Sediments'15, CD-Rom, No. 34; 2015. pp. 1-14

[10] Sakuno Y, Iijima Y, Tsushima K, Kido C. Wave data collection in Indonesian domestic seas using satellite altimeter data. Japan Society of Naval Architects and Ocean Engineers. 2007;**6**:27-33. DOI: 10.2534/jjasnaoe.6.27], 10.2534/jjasnaoe.6.27]

[11] Noshi Y, Uda T, Kobayashi A, Miyahara S, Serizawa M. Beach changes observed in phan Rang City in Southeast Vietnam, 8th Int. Conf. On Asian and Pacific coasts (APAC 2015). Procedia Engineering. 2015;**116**:163-170

[12] Uda T, Onaka S, Serizawa M. Beach erosion downcoast of Pengambengan fishing port in western part of Bali Island. In: 8th Int. Conf. On Asian and Pacific Coasts (APAC 2015), Procedia Engineering. Vol. 116. Elsevier; 2015. pp. 494-501

[13] Miyahara S, Uda T, Serizawa M. Field observation and numerical simulation of barrier island formation as a result of elongation of sand spit and its attachment to opposite shore. Proceedings of 35th Conference on Coastal Engineering, Sediment. 2016;**5**:1-14

[14] Ashton A, Murray AB. High-angle wave instability and emergent shoreline shapes: 1. Modeling of sand waves, flying spits, and capes. Journal of Geophysical Research. 2006;**111**:F04011. DOI: 10.1029/2005JF000422

[15] Uda T, Noshi Y. Recent shoreline changes due to high-angle wave instability along the east coast of Lingayen gulf in the Philippines. Geosciences. 2021;**11**:144. DOI: 10.3390/geosciences11030144

[16] Uda T, Serizawa M. Elongation of a sand spit offshore of groins due to high-angle wave instability. Global Media Journal. 2021;**19**(S7):252

[17] Uda T, Serizawa M, Miyahara S. Morphodynamic Model for Predicting Beach Changes Based on Bagnold's Concept and its Applications. London, UK, London, UK: INTECH; 2018. p. 188

[18] IPCC. Climate Change 2014 Synthesis Report. Intergovernmental Panel on Climatic Change, 151 p; 2014

Chapter 4

A Novel Method of Hydro-Vacuum Dispersion of Metallurgical Melts: Research and Implementation

David Sakhvadze, Gigo Jandieri, Besik Saralidze and Giorgi Sakhvadze

Abstract

In parallel with the gradual expansion of the consumption of powder materials and the increasing demands placed on them, the competition between the producers of powders is continually intensifying. There is no doubt that the future belongs to technologies that provide high productivity and low-cost powders. Consequently, the technology and techniques of powder production need constant revision and modernization. For this goal we have developed and proposed a new method and installation for hydro-vacuum dispersion of melts, the essence of innovation and advantage of which lies in sucking and dispersing the melt in the direction opposite to the action of the force of gravity, under gravity overload 150-200g conditions, where the main work is performed by hydraulic rarefaction resulting from a sharp refraction of direction (on 162-degree angle) and rapid expansion (x10) of a high-pressure water annular flow, with the superimposition of spatial shock- pulsating waves generated in the outer shell of the formed cone-shaped vortex. The device is characterized by high production and low energy costs, while powders - by increased specific surface, improved purity and high activity. The enhanced activity of our powders is due to the formation of non-equilibrium mechanoactivation structural-deformation stresses in them, which leads to the accumulation of excess chemical energy in them. It is justified that the application of the method is also highly effective for dispersing slag melts and obtaining amorphous hardened powdery raw materials with high hydraulic activity, suitable for the production of construction cement. Appropriate recommendations for the industrial implementation of the developed innovative technology have been proposed.

Keywords: powder technology, hydro-vacuum dispersion, high-performance unit, activated metal powder, simulation, industrial implementation

1. Introduction

Among the numerous methods of dispersing metallic melts into functional powders, today, the most widespread in industrial practice is the technology of high-pressure water or gas jet spraying (atomization) of the melt [1, 2]. Nevertheless, due to the difficulty of mastering the technological process and the limited functional capabilities of the equipment (periodic work cycle, technical service difficulty, low productivity, harmful effects on the environment [3, 4]), the issue of industrial unification and large-scale implementation of the mentioned technology remains a problem.

A comprehensive overview analysis of the achievements and problems of further development of modern technologies for the production of powder materials is presented in the study [5].

According to the results of this research especially acute problems are to be solved in 3D printing technology for large-size products and—in the direction of obtaining biocompatible powder materials. Particularly problematic is the need to eliminate the phenomenon of non-uniform temperature distribution in the powder bed when machining large workpieces. This makes it difficult to control the stress distribution and leads to effective deformation of the workpiece. In addition, increasing the size of the moulded part also leads to an increase in the amount of unmelted residual powder. This significantly degrades the mechanical properties of the product and reduces its service life. All this leads to the necessity of an updated approach to both the technologies of additive manufacturing and—methods of obtaining and controlling the quality and properties of the obtained powders. We have to search for new, fundamentally different from conventional technologies methods of obtaining powders with specified sizes, shapes and properties.

Also problematic is the solution of the problem of increasing the efficiency of powders used in the processes of hydromallurgical recovery of minerals from their aqueous solutions, which are produced during mining of the latter [6].

Unlike additive manufacturing, powders used in hydrometallurgy, pyrometallurgy, SHS-metallurgy or other related industries do not need to have a clearly defined spheroidal shape. The main requirement for them is a highly developed surface area (high specific surface area) and a high degree of mechanochemical activation.

Provision of guaranteed one-stage cost-effective production of powders with the noted properties is the primary task of our research. Known and currently used powder production installations cannot directly produce powders with the desired properties. All of them require additional physical, mechanical or chemical treatment.

In addition, the gained practice shows, that due to the peculiarities of the technological processes, it is practically impossible to process melts of different types (of different chemical composition and viscosity) with one plant. In each individual case, the equipment needs to be retooled, to change working nodes and their operating modes. In addition, in the large-scale production of powders (for example, the production of iron powder), operational management of the efficiency of the melt sputtering process, including stabilization of the conditions of formation into individual granules, long-term maintenance of operational properties, elimination of the problem of oxidation, gas saturation and reduction of pyroporosity, has remained an unsolved task until now [7–9]. As a result, the powders obtained by this method are inevitably subjected to additional sorting-mechanical crushing in an inert gas environment, as well as to further thermal, hydrogen-reducing treatment [10, 11].

Similar problems exist in the relatively less common centrifugal dispersion and ultrasonic disintegration processes of atomization metallurgical melts [12–14]. In all cases, the powders require a finishing thermomechanical (activation) treatment [15, 16]. Without this step, the powders remain in a passive (chemically low-activity) phase. Therefore, their direct use in mining and metallurgical, machine-building, construction and other fields is less effective [17–19]. The need to optimize the alloy dispersion process and hydrogen reduction treatment of the obtained powders is especially acute when dispersing active metal alloys such as aluminum, magnesium, Armco Iron, low-carbon alloyed steels, modified cast iron, etc. Due to all of the above, the process of dispersing metallurgical melts into functional powders is considered one of the most high-tech, expensive and difficult production processes, which requires further research and optimization.

Based on the above, the main purpose of the study was the development and implementation of a conceptually new, drastically different methodological approach to the process of dispergation of melts, leading to the direct production of powder particles in a mechanically activated state. To achieve this purpose, it is necessary to solve the task of non-uniaxial volume-shift deformation of particles solidifying during dispergation. Practical provision of obtaining the desired effect is possible by forced acceleration of the melt jet leading to simultaneous gradient linear and angular acceleration, fragmentation and solidification in the field of forces directed opposite to the universal gravitational force. Such an approach to the dispersion process will certainly lead to the formation of solidified melt particles in a non-equilibrium, energetically stressed structural-phase state.

The relevance and significance of the shock-wave effect on the dispersed jet is also emphasized in the studies [20, 21], which found that the liquid metal droplet undergoes breakup through the shear-induced entrainment mode for the studied range of Weber number values (We ≈ 400–8000). The prevailing mechanism is explained based on the relative dominance of droplet deformation and Kelvin-Helmholtz wave formation.

We have established that such a special condition necessary for obtaining the desired effect is the initiation of vertical suction and hydromechanical dispersion of the melt, carried out with the help of a specially formed for this purpose lifting force of a high-speed annular water flow with a rarefied (hydro vacuum) core. The new method and device constructed in this case should provide high productivity, technological flexibility and versatility, where the problem of eliminating the need for additional chemical and physical-mechanical treatment of the resulting powders should also be solved.

2. Research methodology and materials

As a rational solution to the above problem, we chose the way of improving the design of the pilot unit [22] and expanding the functional capabilities of the innovative method of hydrovacuum dispersion of alloys [23], specially developed by us earlier for the above purpose. The general arrangement of the new dispersion unit and its main process juncture is illustrated in **Figure 1**.

The main technological unit (*Hydro-vacuum Dispergator*) contains: the body/housing 1 with a lateral high-pressure water tangentially delivery elbow pipe 2; the Venturi tube of relatively small dimeter and length 3 concentrically mounted in the body 1, with bottom position of the inlet duct 3′ and the circular collector channel 4 formed between the inner surface of the body/housing 1 formed and shell of Ventura tube 3, coupled with the high-pressure water delivery lateral elbow pipe 2; the semi-toroidal internal (convergent-divergent) confuser 5 formed through the lower end of the Venturi tube 3–3′ and the collector channel 4 in the extreme lower peripheral part of the body/housing 1; the suction nozzle 6 axially placed from its inner surface at a (l) longitudinal and (h) transverse distance with a cylindrical perforation 6′ (d);

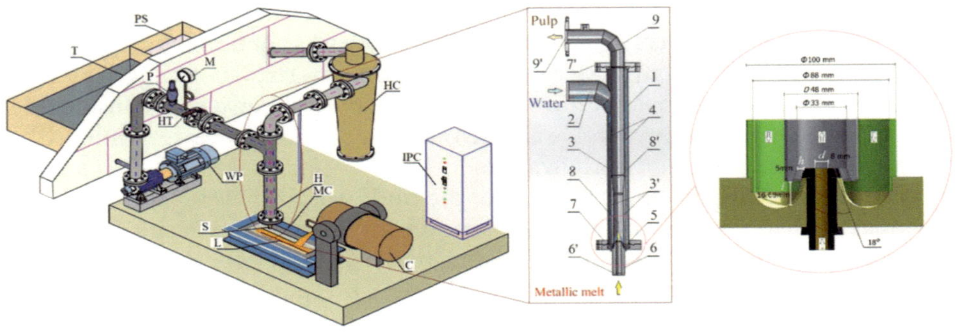

Figure 1.
Innovative installation of hydrovacuum dispersion of metallurgical melts: H—hydro-vacuum dispergator; WP—electric drive of water pump; M—manometer; P—pressure regulator; HT—high-pressure water tap; IPC—inverter for smooth pressure control; S—suction nozzle; C—casting ladle; M—mold for casting; L—liquid melt; HC—hydrocyclone; PS—pulp precipitation sector; T—tank for purified water.

lower 7 and upper 7′ flat toroidal flanges for sealing the water-injection hydraulic system (body 1, annular duct 4); in the converging duct 3′ of the Venturi tube 3, the dispersion chamber 8, formed on the top of the sucker 6 with the diffuser 8′; the pulp discharger 9 within the diffuser 8′ flexibly connected to the upper end of the pipe 3, and the guiding channel 9′ in its hydrocycline-type collector-settler.

For research and technological improvement of the design developed by us, the simulation modeling method was applied. In particular—the computer software module Solidworks Flow Simulation [24] based on the laws of theoretical hydrodynamics and hydromechanics was used for the simulation modeling of the device. When modeling, the flow rate (velocity) of the water supplied to create a hydro-vacuum in the device was used as a variable parameter, the range of variation of which was within the limits of 0.015–0.029 $m^3\ s^{-1}$. During the calculations, variation of the geometrical parameters (l, h) of the semi-toroidal confuser (5) forming a conical vortex from the water supplied at high pressure from the circular collector channel (4) of the dispersion-diffusion chamber (8–8′) of the body (1), from the (initial) values l = 16.69, h = 5 mm, within ±50% deviation limits, obtained when making design solution. At the same time, the pressure of the water supplied in the circular collector channel (4) of the device was varied in the range of 4–23 bar. Through the comparative analysis of the modeling results carried out as a result of the mentioned approach, the optimal values of the construction and technological parameters necessary and sufficient to ensure the desired theoretical value of hydro-vacuum (−1 bar) were determined.

In order to evaluate the accuracy and effectiveness of the data obtained as a result of the modeling, standard methods and tools of hydrometry were used for the experimental research, including a digital water pressure measuring manometer with an electronic transmitter PTL-25A (measurement range 0–25 bar) and a vacuum meter Leybold THERMOVAC TTR 91 (measurement range $5.10^{-4} \div 1000$ microbar).

For recording vibrations of the hydro-vacuum dispersion chamber, a digital gyroscope mounted on the device, a noise meter (Integrating Sound Level Meter) for recording the acoustic spectrum, including the ultrasonic range of acoustic noise, liquid metal temperature measuring contact thermocouples with digital galvanometers, etc. were also used. Accumulation and processing of experimental data was carried out online, directly during the conduct of experimental studies, through the use of a special experimental computer platform Grab ExP.

Experimental studies were conducted, alternately, both using a high-purity aluminum of the 1000 series, aluminum scrap, secondary gray cast iron and its slag melts. An induction furnace Termolit ICMEF-0.06 was used for melting. Structural and metallographic studies were performed using a metallographic microscope Neophot 32 (Carl Zeiss).

To evaluate the quality of the activity of the obtained aluminum powder, a special tool—a FANN-432 model manometric calcimeter [25], with an automated magnetic stirrer and an automatic sodium hydroxide supply function was used, which led to the rapid removal of the oxide film from the surface of the investigated particles and an increase in the rate of hydrogen release from dissociated water molecules.

To evaluate the activity of iron powder obtained from the melt of secondary cast iron, the method of analyzing the cementation (extraction) ability of copper from acid sulfide solutions was used [26]. In turn, to assess the hydraulic activity of the obtained slag granulate, the method of lime (CaO) absorption was applied. For this purpose, an aqueous solution of lime ($Ca(OH)_2$) with a density of 1.0 g/cm^3 was prepared. The change in CaO concentration was monitored at intervals of 24 h, by titrimetric method with 1 H hydrochloric acid solution, according to GOST 22688-2018.

Structural and metallographic studies of the obtained powder particles were carried out using both metallographic optical microscope "Neophot 32 (Carl Zeiss)" and autoemission scanning electron microscope "JSM-7800F" and X-ray diffractometer "DRON-7".

3. Technological features and advantages of the hydro-vacuum dispersion process

In contrast to the traditional methods of production of functional powders of metallurgical melts for various purposes, the essence of the innovativeness and advantage of our proposed technology lies in the suction-dispersion of metallurgical melts in the opposite direction of the force of gravity, under 150–200 g gravity overload conditions, where the main work performs the rarefaction formed in the vortex core of the conical profile formed as a result of the rapid expansion of the annular flow delivered by high pressure and the sharp 162° change of the flow direction in a closed space (hydraulic vacuum). The shear forces (tensile stresses) vertically, bottom-up created on the surface of the cylindrical jet of melt absorbed in the vacuum action zone, inclined to each other by 36°, and directed upwards by 72° from the horizontal plane, create fundamentally different and hydrodynamically optimized conditions of melt dispersion. At the same time, intense processes of hydraulic friction, hydromechanical adhesion and capillary disintegration take place, where, in contrast to traditional technologies, the impact of hydrodynamic flows of water with increasing angular acceleration on the particles (droplets) taken from the dispersed melt continues until they solidify and enter a special collector. This additionally conditions large-scale 3D twisting-deformation and intense micro-volume cavitation-impulse (ultrasonic acoustic) treatment of the formed particles, thus achieving the irreversible effect of their microstructural segregation, gradient migration and tension (mechanoactivation). Consequently, targeted powders are obtained with improved purity, increased surface area and activity.

The process of vertical suction and dispersion of the metallic melt in **Figure 2** is represented as a hydro-mechanical scheme.

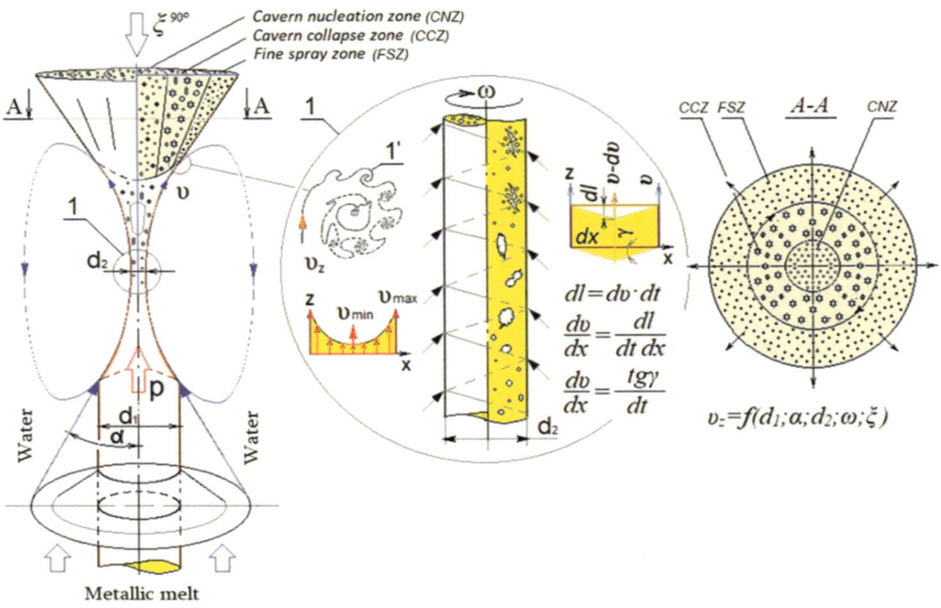

Figure 2.
Hydro-mechanical scheme of hydro-vacuum dispersing process: υ—linear velocity of displacement of melt particles, m/s; α—angle of attack of injected water on the suction jet, deg; d_1—initial diameter of the suction jet, mm; d_2—diameter of the jet after vortex-stretching influence, mm; ω—angular velocity of rotation, rad/s; ξ—value of hydraulic resistance, kg/m^3.

The vertically sucked liquid metallic jet, extending through the zone of hydraulically rarefied funnel formed in the core of the cone-shaped water jet shell injected from under the semi-toroidal annular confuser (5), in the process of axial displacement and exit from the compression zone, under the influence of force hydro-mechanical impact narrows in cross-section ($d_1 > d_2$), accelerates, and then, due to elastic forces of acceleration expands again, accelerates, and in the consequence of excited powerful centrifugal-tangential tensile forces destroys, splashing in the dispersion chamber (8) with small shapeless drops. The process proceeds so quickly that the possibility of vapor accumulation and excessive pressure rise is completely excluded. At the same time, vapor shells formed around the surfaces of high-temperature metal droplets (with a critical temperature of 375°C [27]), in addition to intensive heat exchange and cooling effect, increase the potential energy of repulsion, contributing to their quick removal from the zone of rapid hydrodynamic interaction. It is noteworthy that both before entering and after leaving the zone of primary radial compression, force stretching and vortex twisting (**Figure 2**, Section 1), in the body of the metal jet, drawn vertically directed cone-shaped core of the funnel, under the influence of the formed dynamic vacuum and shock waves of rarefaction, bubble cavitation clusters (caverns) with the effect of pseudo-boiling are formed. These clusters progressively expand (grow) as the absorbed metal is released from the zone of hydrodynamic compression, stretching and twisting. It is in the marked zones of pseudo-boiling, where the density and strength of the liquid metal weakens the most, that centers of disintegration appear, supplied in the dispersion chamber (8) of the liquid metal jet. Some of the gases released after the disintegration of the liquid metal jet (**Figure 2**, zone FSZ) were carried away with water, and the other part, most

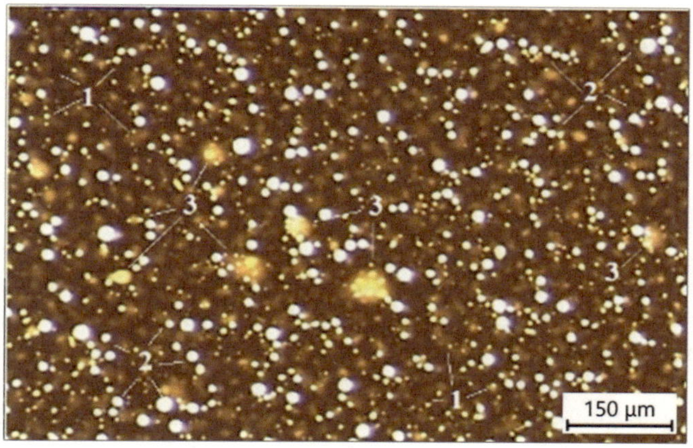

Figure 3.
Nucleation (1), growth (2) and collapse (3) of cavitation cavities in the process of hydrovacuum dispersing.

likely, together with the shells of water vapor formed on the surfaces of the particles, formed new cavitation bubbles. As the marked vapor-gas mixture cooled and condensed, and, accordingly, the internal pressure of the shell dropped, which occurred in micro- and nanosecond intervals, these cavitation bubbles collapsed and generated cumulative micro explosions coinciding in the direction of its flow vector. Due to the high dispersion speeds (water supply speed 250–400 m/s) and, consequently, the high intensity of cavitation detonations, the frequency of these micro-explosions was in the ultrasonic range. Photographs showing the formation, growth, and collapse of cavitation caverns are presented in **Figure 3**, obtained as a result of high-speed photofixation (exposure rate 10^5 frames/s).

The above-mentioned cumulative impact effect, caused by the phenomenon of cavern collapse, clearly manifested itself in **Figure 4**, which shows an illustration of the surface of a powder particle with a cavitation ablation crater, and with subsequently formed satellite nano-sized spheroidal particles.

From the point of view of explaining the hydromechanical laws of the process of angular twisting of a dispersed jet and shear deformation of individual droplets emitted from it, it is necessary to emphasize that, based on the laws of hydrodynamics and the

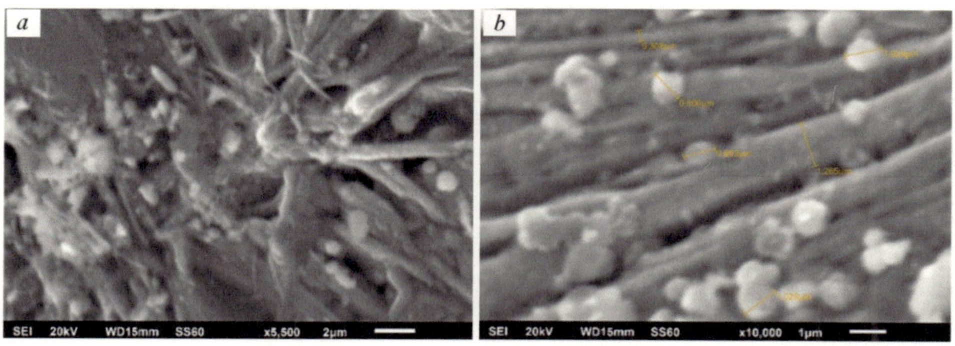

Figure 4.
The surface of a powder particle with a cavitation ablation crater (a), and with subsequently formed satellite spheroidal nanosized particles (b).

properties of Newtonian liquids, in the process of hydromechanical capture and acceleration of the melt, in the peripheral layers of the latter, due to differences in displacement velocities across the jet cross-section (the central zone always lags behind the peripheral zone, $\upsilon_{min} < \upsilon_{max}$), the initial conventional rectangular microparticle of the dispersed melt will always deform linearly and turn into a parallelogram. In the diagram (**Figure 2**, Section 1), this deformation is indicated by a shift dl. The segment dl characterizes the magnitude of the deformation during the time dt, i.e. $dl = d\upsilon \cdot dt$, then the velocity gradient $d\upsilon/dx = dl/dt \cdot dx$ but $dl/dx = tg\gamma$, then $d\upsilon/dx = tg\gamma/dt$. Therefore, the transverse velocity gradient is the relative shear/displacement strain rate. Thus, the tangential stresses important from the point of view of the fragmentation of the sucked melt jet depend linearly on the relative strain rate, which means that, unlike the deformation of solids, the magnitudes of the tangential stresses and the degree of associated linear and angular shear, in this case, mainly depend on the hydrodynamic injection power, the angle of attack α, the depth of the associated rarefaction and the area of its effect, rather than on the simple linear velocity of the water flow υ. The formation of a hydraulic structure of a cone-shaped dynamic vortex with an evacuated funnel, adjustable in power and geometric parameters, providing vertical (axial) intake and dispersion of the target melt through the force action of rarefaction shock waves with accompanying cavitation pseudo-boiling fronts, can be considered one of the main distinctive technological features of the process of obtaining metal powders by the method hydro-vacuum dispersion.

The mentioned phenomenon provides obtaining powder particles with fibrous, non-equilibrium structurally tense (i.e. mechanically activated) state, which is a result of flattening and asymmetric twisting of drops of liquid metallic breaking away from the liquid metal sucked in the dispergator under conditions of volumetric impact by shock-pulsating waves of hydraulic rarefaction. A schematic of this process with the formation of the Kelvin-Helmholtz wave effect is depicted in longitudinal Section 1′ in **Figure 2**. This feature can be considered as an essential distinguishing sign of particles obtained by hydrovacuum dispersion of melts. This feature is clearly illustrated in **Figure 5**, which shows the characteristic morphology and microstructure of a particle obtained by hydrovacuum dispersion of liquid cast iron.

The technological system implementing the mentioned process (operating plant, water circulation mechanism, pulp separator, etc.) is distinguished by its compactness and mobility. It can be used practically in metallurgical production of any scale for large-tonnage production of activated, functional powder from any type of molten metal.

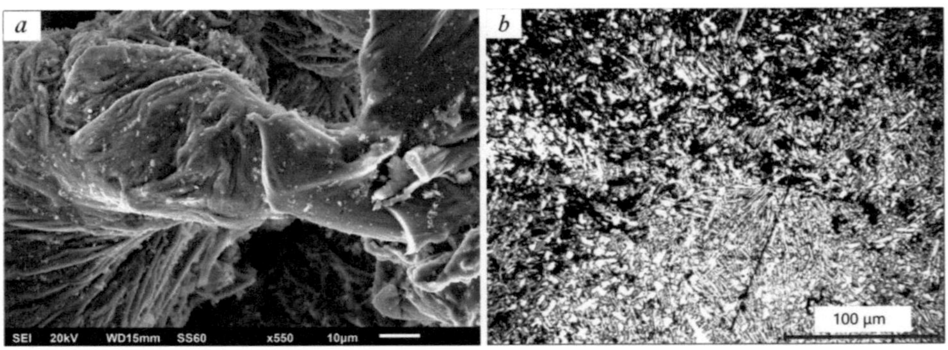

Figure 5.
Surface morphology (a) and microstructure (b) of iron powder particles obtained by hydrovacuum dispersion of melts.

In order to regulate the granulometric composition of the powder particles, the device also provides for the regulation of the rarefaction formed in the core of the vortex of the melt suction cone profile, both by controlling the speed and pressure of the water supplied to the device, and also by changing the vortex profile (shape and height), which is achieved by a semi-toroidal confusor that vertically sprays water supplied by pressure in the body of the device by adjusting the section (dimensions l, h) of the confuser (convergent-divergent) nozzle (5).

Implementation of such a fundamentally different controlled process provides the possibility of continuous production of absolutely homogenized pulp, and therefore, relatively homogeneous, highly activated functional powders.

The proposed technology is environmentally safe. No smoke, steam or dust is released during the work process, because the water circulating in the technological system itself performs the function of both an air filter and a steam condenser. The device operates quietly and without significant vibrations, is compact and easy to install (installation is not related to significant reconstruction of the smelting plants or foundry area). The technical water used in the working area circulates in its own pipe system and also does not pollute the environment. The technology is energy-efficient, eliminates high-temperature heat treatment of the received powders in a regenerating environment. Therefore, it is highly profitable and cost-effective.

4. Research results and discussion

The results of simulation modeling and parallel experimental research (**Figure 6**) showed us that in order to operate the device with the maximum possible efficiency, which was expressed in the condition of providing technical −1 bar power vacuum with minimum expenses (water pressure), it is necessary to reduce the parameters h and l of the semi-toroidal confuser (**Figure 1**, position (5)) to the optimal values obtained by the simulation model. These values are: h_{opt} = 3.24 mm, l_{opt} = 8.75 mm.

It was determined that, as a result of the optimization of the mentioned parameters, in order to obtain −1 bar rarefaction in the hydro-vacuum dispersion chamber (**Figure 1**, position 8), in exchange for the supply of water at a rate of 0.029 $m^3 s^{-1}$ into the circular collector channel (4) with a pressure of 16–20 bar used by us so far it will be enough to develop a speed of 0.022 $m^3 s^{-1}$ and a pressure of 13.2 bar. The effectiveness of the mentioned solution is clearly proven by the results of the report on the acceleration rates of the melt particles absorbed/dispersed under the influence of the pressure drop and created rarefaction in the confuser-diffuser pair (5–8), which is presented in **Figure 7** in the form of diagrams.

From the diagrams presented, it can be seen that the reduction of the height (h) of the confuser (5) by 35% and the thickness (l) by 47% increases the pressure drop rate in the mentioned transition zone from 11 to 8 bars and increases acceleration of the water-carried metal particles from 3.0–3.5 to 4.1–4.3 × 10^2 m s^{-1}, which also leads to a change in the amplitude-frequency characteristic of shock waves generated in the device. This is one of the key factors in terms of improving the intensity of melt dispersion and the kinetics of mechanoactivation of the formed powder particles.

From the diagrams showing the results of the reporting data, it can also be seen that as a result of the optimization of the geometrical parameters of the confuser (5) in the diffuser part of the hydro-vacuum device 8–8′, the amplitude of the shock-wave impact generated on average decreases from 0.35 to 0.065 units. On the other hand, it is observed that the frequency decreases by 40% after reaching the height of

Figure 6.
Simulation model of hydro-vacuum dispersing unit operation and prototype in industrial conditions.

120 mm, and in the optimized case, on the contrary, it increases sharply after reaching the height of 180 mm. In addition, it can be seen from the diagrams that if in the first case the zone of active dispersions is located in the height range of 25–180 mm, as a result of parametric optimization, the height of this zone increases to 250 mm. This also indicates the activation of cavitation processes and the strengthening of shock waves accompanying the rarefaction front.

By checking the adequacy of the obtained results, which was carried out by modernizing the structural elements of the device in accordance with the simulation model and a new trial-experimental investigation, we were convinced of the high (≈95%) accuracy and efficiency of the latter. This is clearly confirmed by the graphic material constructed as a result of the experimental research data presented in **Figure 8**, where it can be seen that as a result of the parametric optimization of the device and process, similar to the simulation model, the amplitude of the vibrations of the dispersion chamber decreases, and the frequency increases, so that the gravitational overload, that is, the average rate of the vacuum suction-acceleration of the dispersed particles decreases relatively insignificantly, by 0.004 m s^{-2}.

The above is further supported by the data recorded by the digital audio recorder and the acoustic spectrograms obtained as a result of their processing in **Figures 9** and **10**. In particular, it can be seen from these spectrograms that after parametric optimization of the hydro-vacuum device, the power of the noise emitted during its operation, i.e. the density of its spectrum, decreases at the expense of reducing the amplitude of acoustic waves, i.e. the kinetic energy. This leads to the ability to transfer the potential energy accumulated in these waves to a greater distance (height), and this is very important in terms of increasing the efficiency

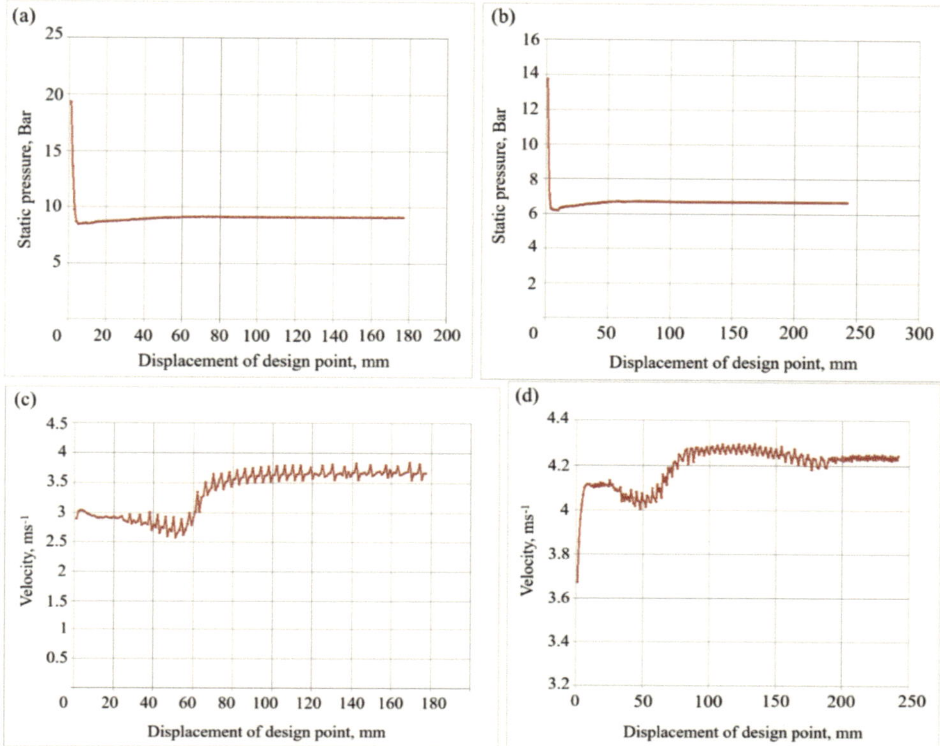

Figure 7.
Results of the simulation study: Indicators of pressure drop and velocity increase of the water supplied to the dispersion chamber in relation to its vertical location, before optimization ((a) and (c)) and after optimization ((b) and (d)) of the dispersing process.

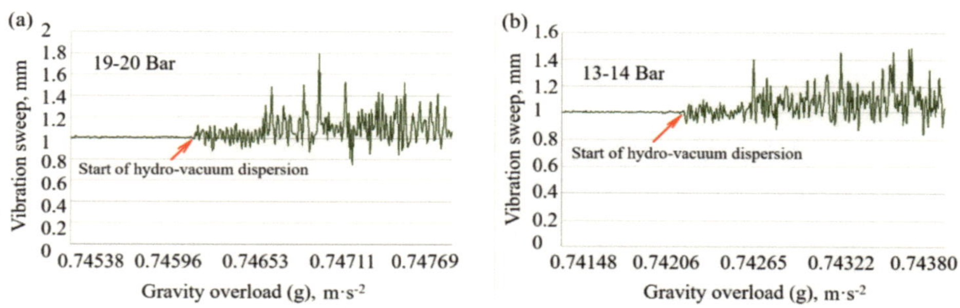

Figure 8.
Vibration of the dispersion chamber of the hydro-vacuum unit before (indicated by the red arrow) and after the working process: (a) vibration's sweep, mm and overload dynamics before optimization of operating parameters and water pressure, (b) after optimization.

of cavitation currents acting on dispersed particles, i.e., secondary mechanoactivators. The said is also quite similar to the results of the above-mentioned simulation modeling (**Figure 5(d)**) and supports it.

In order to compare the hydromechanical processes generated in the dispersion chamber of the device, in particular, the model and actual profiles of the force hydrodynamic currents, during the experimental work, the procedure of artificial freezing

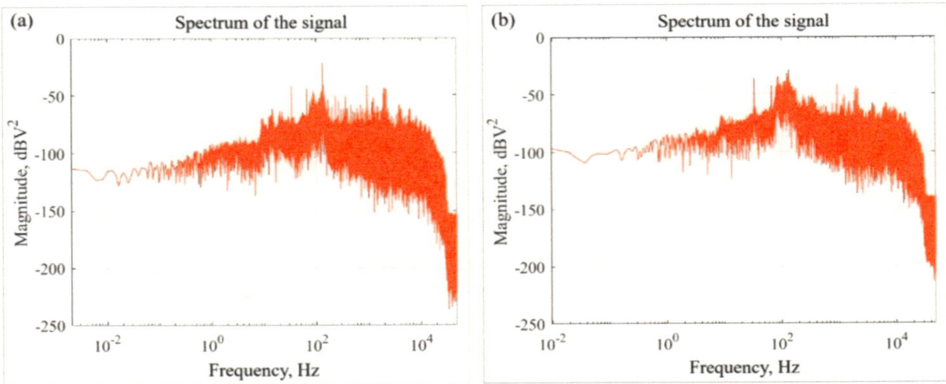

Figure 9.
Amplitude spectrum of the audio signal before and after pumping the melt into the dispersion chamber: (a) spectrum before optimizing the geometrical parameters of the converging duct (5); (b) after optimizing.

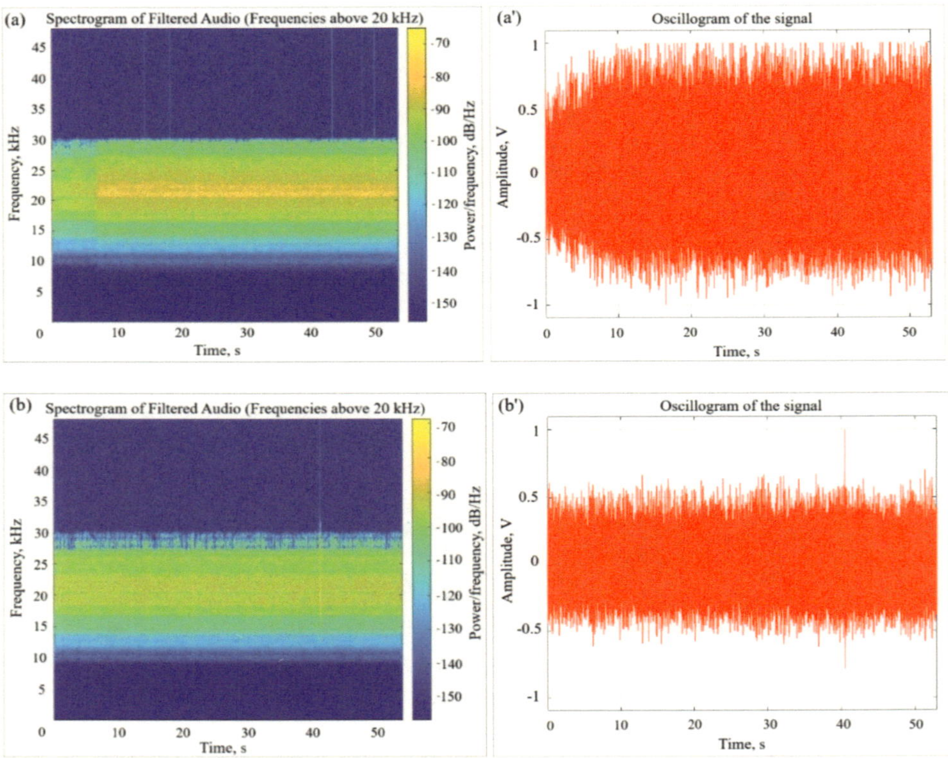

Figure 10.
Acoustic spectrum of the hydro-vacuum dispersion process in the established operating mode of the equipment: (a, a') before optimization and (b, b') after it.

of the jet sucked into the device was carried out, which was carried out by means of a gentle reduction of the water pressure and braking of the dispersion process.

Based on the imprint (**Figure 11**) left on the peripheral layers of the cylindrical flow of molten metal absorbed by the hydrodynamic shell of the vortex of water

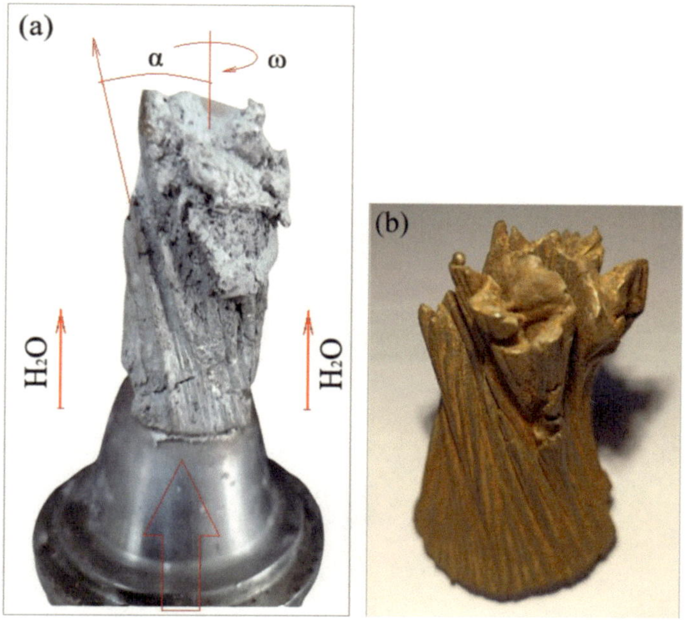

Figure 11.
Effect of hydromechanical action on the suction metal flow: (a) suction profile for aluminum and (b) the same for gray cast iron.

sucked into the disperser and conically ejected from the confuser (5), there is no doubt that the ejected water shell, passing through the 36° inclined skirts of the conical cap of the suction nozzle (6), experiences angular acceleration corresponding to the direction of negative rotation of the water tangentially delivered from the tube (2) in the circular collector channel (4), identical to the forces shown in the graphical simulation model (**Figure 2**).

It is noteworthy that according to the existing imprints, the actual indicator of the inclination of the vector of the angular movement of the hydrodynamic vortex of water was the value provided by the design $\alpha \approx 18 \pm 0.5$ deg, which indicates the operation of the device in conditions as close as possible to the optimal ones. Deviation from the optimal values of the technological regimes for the larger value (water pressure 13.2 bar, velocity 0.022 $m^3 s^{-1}$) led to a decrease in the mentioned angle, and to a small value, on the contrary, to an increase.

The study of the granulometric composition of the obtained powders, the results of which, in turn, are presented in the form of histograms illustrated in **Figure 9**, showed that the dominant fraction of particles obtained after the optimization of the hydro-vacuum dispersion process is 100–150 μm, which is on average 60–80 μm less than the fraction formed before the optimization of the process (**Figure 12**). In addition, its specific share has increased from 10–15% to 25–34%. This effect is also clearly illustrated by the micro-images shown in **Figure 13**.

The laboratory analysis performed on the manometric calcimeter determined that the reduction of the dispersity of the obtained aluminum powder particles led to an increase in the content of active particles from 95–96% to 99%, which is explained by an increase in the specific surface of the latter and the degree of mechanoactivation, that is, in the concentration of the hidden physico-chemical energy accumulated in them.

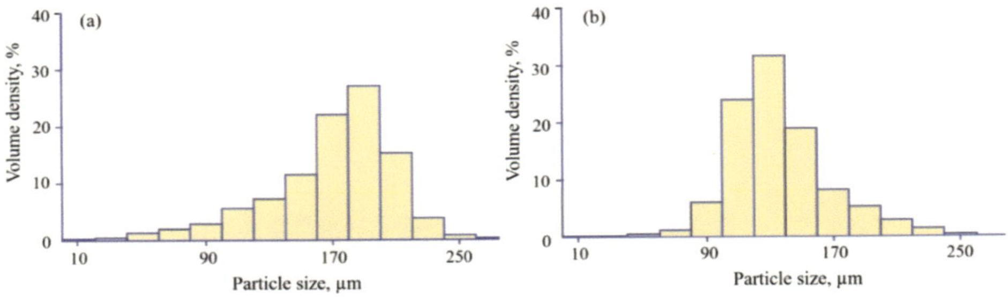

Figure 12.
Granulometric composition of obtained powders before (a) and after (b) optimization of the hydro-vacuum dispersion process.

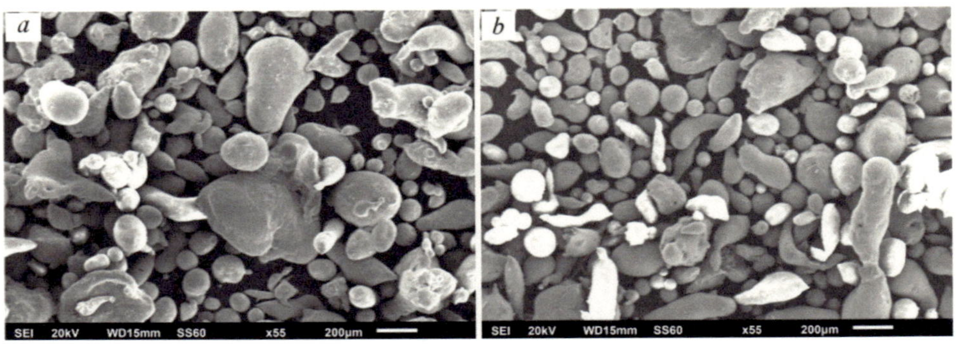

Figure 13.
Cast iron powder microparticles before (a) and after (b) optimization of the hydro-vacuum dispergation process.

Experimental studies also revealed the need to manage the technological parameters having a significant influence on the granulometric composition of powders, such as: the initial temperature of the melt and the speed of its absorption in the disperser. It is significant that the diameter of the inner cylindrical channel of the suction nozzle is the key factor in regulating the speed of melt absorption, i.e. dispersion cost. It was determined experimentally that in the case of hydro-vacuum dispersion of aluminum, the rational weight ratio "liquid metal: water" is 1:20 ÷ 1:23. Therefore, in the case of a water velocity of 0.022 $m^3\ s^{-1}$, the optimum speed of aluminum supply to the dispersion chamber shall be 0.0011–0.00095 $m^3\ s^{-1}$, which in mass units corresponds to a speed of 2.5–3 kg s^{-1}.

When dispersing liquid cast iron, the rational weight ratio "liquid metal: water" is 1:50–1:55; the optimal vacuum depth is −0.85 ÷ −0.91 bar, achieved at the pressure of injected water 12–16 bar.

The operational management of the process is carried out through a programmed operating computer/laptop, where the function of the supporting signal is performed by the operationally controlled granulometric composition of the produced powder. In case of deviation from the latter's target indicators, a control impact signal is generated by means of a special portable express analyzer integrated with a computer with a special controller, based on which the computer program using the MATLAB simulation modeling database automatically makes a decision to manage the dispersion process, in particular, the suction speed of the molten metal, by appropriate adjustment of the pressure and consumption of the water tangentially entering in the circular collector channel (4) and semi-toroidal convergent-divergent confuser (5).

The results of the analysis of the chemical activity of the iron powder we obtain clearly showed the advantage of our method of hydrovacuum dispersion of melts. In contrast to conventional powdered iron (e.g. CMS Magnetics brand, 200 microns, Fe 99.95%), with the help of which copper is extracted from sulfate solutions and quarry wastewater from the mining and enrichment of copper ores of the Kazreti deposit (Georgia) in the form of its divalent hydroxide $Cu(OH)_2$, which requires subsequent energy-intensive reduction firing in a hydrogen environment, our powder, due to the presence of an amorphous phase, predominantly containing γ modification of iron (Fe_γ) [28], directly ensured the production of particles of pure elemental copper. The sequence of chemical reactions taking place was as follows:

$$Fe_{\tilde{a}} + Fe_3C + 4CuSO_4 + 4H_2O \rightarrow 4FeSO_4 + 2Cu + 2Cu(OH)_2 + 2H_2 + C \quad (1)$$

$$Fe_{\tilde{a}} + Fe_3C + 6H_2SO_4 \rightarrow 2Fe_2(SO_4)_3 + C + 6H_2 \quad (2)$$

$$Fe_2(SO_4)_3 + 3Cu(OH)_2 \rightarrow 3CuSO_4 + 2Fe(OH)_3 \quad (3)$$

$$CuSO_4 + Fe_{\tilde{a}} \rightarrow FeSO_4 + Cu \quad (4)$$

In contrast to the traditional cementation process [26], here the iron consumption is reduced due to the formation of Fe^{3+} trivalent iron ions ($Fe_2(SO_4)_3$) active for copper extraction. The need for sulfuric acid is also reduced because, in contrast to the compared process, there is no formation of free water molecules neutralizing the set acidity. As a result, it leads to the possibility of maintaining the acidity at a less aggressive level (pH 2.0–2.5 instead of pH 1.0), gives the opportunity to reduce the molar ratio of copper-iron M(Cu):M(Fe) to 1.0:1.1, which leads to savings of iron powder consumption by 30–35% to 2.5–3 kg/m^3.

As a result of experimental studies, it was also found that the use of hydrovacuum dispersion method is also highly effective for dispersing slag melts and obtaining from them amorphous solidified (glassy) powdery raw materials with high hydraulic activity, suitable for the production of construction (Portland) cement [29]. In the case of hydrovacuum dispersion of slag, the rational weight ratio "slag: water" was 1:18.

Figure 14 shows that the surface of the grains is smooth and vitreous, without crystal roughness, the grains do not contain pores and metallic inclusions. Comparatively small granules consist of smoother and transparent, vitreous formations.

XRD spectral analysis (**Figure 14**(2)) showed that the XRD diffractogram does not contain clearly distinguished diffraction peaks of the crystalline component, which of course indicates the absence of such phases, accordingly, it can be stated that the structure is mainly composed of vitreous, hydrous active, amorphous phase.

The results of the analysis of hydraulic activity of the obtained slag granules showed that on average their activity on CaO absorption is 650–750 mg/g, which is almost twice higher than that of slag granulated by traditional water technologies [21, 30, 31]. The slag granulated in this way, in addition to its use in the production of construction and ceramic materials [32], can also be successfully used as a substitute for mineral abrasives applied for sandblasting of parts for various purposes [33].

Figure 14.
Morphology (1), EDS and XRD spectrograms (2) of granules obtained by hydro-vacuum granulation of gray cast iron slag [29].

The proposed method of hydrovacuum dispersion of metallurgical alloys and the device for its realization on the basis of the above experimental studies is implemented and started up at the production site of "Geo Enterprise" LLC. [34].

5. Conclusions

A systematic analysis of the above results allows us to conclude that:

1. A new method for dispersing metallurgical melts has been developed, the essence of which is to initiate the process of forced vertical vacuum suction of the liquid melt in the form of a thin jet and its droplet fragmentation by gradient (layer-by-layer) acceleration in a force field of hydromechanical action. The source of the suction effect is the rarefied core of the water vortex formed in the central zone of the dispersion chamber;

2. The process is dynamically controlled and can be carried out continuously. The pulp formed with water cools instantly without emitting steam, smoke or dust;

3. The productivity of the proposed method and dispersing unit is on average 10 times higher than that of conventional atomization installations of identical dimensions. This increase in productivity will undoubtedly lead to a significant reduction in the production cost of the resulting powders. In addition, dynamic vacuumization of metal flow and hence additional possibility of its physical purification from harmful non-metallic and gas impurities can lead to the possibility of using cheap secondary metals and alloys as a starting material instead of currently used pure metals;

4. A schematic hydromechanical and computer simulation modeling of the operating features of the dispersion chamber was carried out. A comparative analysis of the data obtained by modeling with experimental data was performed. It is determined that the further improvement of the simulation model and the consideration of parameters of such essential importance as the initial temperature of the melt, its viscosity and the speed of its suction in the dispersion chamber, will give us the opportunity to identify the necessary technical conditions for the optimization of hydro-vacuum dispersion regimes for the free-flowing melt of practically any type and technological complexity without conducting preliminary experimental studies;

5. Morphological and physio-chemical studies of the obtained powders have shown their advantage over analogs in terms of mechanochemical activity, which is achieved by the peculiarity of the proposed method of hydromechanical impact on the surface of the hardenable particles unevenly three-dimensional shock waves of rarefaction and pulsating cavitation detonations;

6. It has been established that the method and unit of hydro-vacuum dispersion are universal and can also be effectively used for the treatment of slag melts. It is shown that the high degree of structural amorphization and glass transition of the obtained slag microgranules gives them high hydraulic activity suitable for the production of Portland cement;

7. From the viewpoint of industrial implementation of the developed unit, the optimal operating modes of the hydro-vacuum device for dispersing metal melts are: water pressure—13.2 bar, water supply rate—0.022 $m^3 s^{-1}$, thickness of the outlet neck of the annular confuser for the reverse ejection (on 162°) of water—3.24 mm, the suction speed of the melt—2.5–3 kg s^{-1}, which, when working on aluminum melts, is provided by the suction cylinder nozzle with the inner channel diameter of 8 mm. In the case of dispersing liquid cast iron, the rational diameter of the suction cylinder nozzle is 4.5 mm, and in the case of granulation of slag melts—15 mm.

Acknowledgements

The authors acknowledge funding support from the Shota Rustaveli National Science Foundation of Georgia, Grant no. AR-22-1495.

For technical support, the authors also acknowledge the directors of the R. Dvali Institute of Mechanics of Machines and "Geo Eenterprise" LLC, Mr. Tamaz Natriashvili, (Academician of the National Academy of Sciences of Georgia) and Mr. Zurab Magradze.

For support during the modeling and technical measurement phase, special thanks to Anzor Kuparadze and Nata Sulavkvelidze, managers of G3D Ltd.

Conflict of interest

The authors declare no conflict of interest.

Author details

David Sakhvadze[1], Gigo Jandieri[2*], Besik Saralidze[3] and Giorgi Sakhvadze[4]

1 R. Dvali Institute of Machine Mechanics, Tbilisi, Georgia

2 Metallurgical Engineering and Consulting Ltd, Tbilisi, Georgia

3 F. Tavadze Metallurgy and Materials Science Institute, Tbilisi, Georgia

4 Metal Powder Ltd., Tbilisi, Georgia

*Address all correspondence to: gigo.jandieri@gmail.com

References

[1] Dunkley JJ. Metal powder atomisation methods for modern manufacturing. Advantages, limitations and new applications for high value powder manufacturing techniques. Johnson Matthey Technology Review. 2019;**63**(3):226. DOI: 10.1595/205651319X15583434137356

[2] Wallner S. Powder production technologies. Berg Huettenmaenn Monatsh. 2019;**164**:108-111. DOI: 10.1007/s00501-019-0832-2

[3] Korobeinikov I, Perminov A, Dubberstein T, Volkova O. Modification of liquid steel viscosity and surface tension for inert gas atomization of metal powder. Metals. 2021;**11**:521. DOI: 10.3390/met11030521

[4] Fredriksson C. Sustainability of metal powder additive manufacturing. Procedia Manufacturing. 2019;**33**:139-144. DOI: 10.1016/j.promfg.2019.04.018

[5] Shanthar R, Chen K, Abeykoon C. Powder-based additive manufacturing: A critical review of materials, methods, opportunities, and challenges. Advanced Engineering Materials. 2023:Art 2300375. DOI: 10.1002/adem.202300375

[6] Binnemans K, Jones PT. The twelve principles of circular hydrometallurgy. Journal of Sustainable Metallurgy. 2023;**9**:1-25. DOI: 10.1007/s40831-022-00636-3

[7] Powell D, Rennie A, Geekie L, Burns N. Understanding powder degradation in metal additive manufacturing to allow the upcycling of recycled powders. Journal of Cleaner Production. 2020;**268**:122077. DOI: 10.1016/j.jclepro.2020.122077

[8] Grubbs J, Sousa BC, Cote D. Exploration of the effects of metallic powder handling and storage conditions on flowability and moisture content for additive manufacturing applications. Metals. 2022;**12**:603. DOI: 10.3390/met12040603

[9] Vignes A, Krietsch A, Dufaud O, Santandréa A, Perrin L, Bouillard J. Course of explosion behaviour of metallic powders – From micron to nanosize. Journal of Hazardous Materials. 2019;**379**:120767. DOI: 10.1016/j.jhazmat.2019.120767

[10] Baranovskiy A, Pribytkov G, Krinitcyn M, Homyakov V, Dankovcev G. Mechanical activation of self-propagating high temperature synthesis in titanium, carbon black and iron-based alloy powder mixtures. AIP Conference Proceedings. 2019;**2167**(1):20031. DOI: 10.1063/1.5131898

[11] Liu X, Zhang X, Li J, Zhu Q, Deen N, TangY. Regeneration of iron fuel in fluidized beds part II: Reduction experiments. Powder Technology. 2023;**420**:118183. DOI: 10.1016/j.powtec.2022.118183

[12] Zverovshchikov A, Bolshakov G. Simulation of the kinematics and gas dynamics of the centrifugal dispersion stand. Journal of Physics: Conference Series. 2021;**2094**:042046. DOI: 10.1088/1742-6596/2094/4/042046

[13] Cegarra S, Pijuan J, Riera M. Cooling rate modeling and evaluation during centrifugal atomization process. Journal of Manufacturing and Materials Processing. 2023;**7**:112. DOI: 10.3390/jmmp7030112

[14] Kustron P, Korzeniowski M, Sajbura A, Piwowarczyk T, Kaczynski P,

Sokolowski P. Development of high-power ultrasonic system dedicated to metal powder atomization. Applied Sciences. 2023;**13**(15):8984. DOI: 10.3390/app13158984

[15] Politano O, Fourmont A, Le GS, et al. Mechanical activation of metallic powders in planetary ball mills: Multi-scale modeling and experimental observation. IOP Conference Series: Materials Science and Engineering. 2019;**558**:012034. DOI: 10.1088/1757-899X/558/1/012034

[16] Le GS, Bernar F. The mechanical activation of metallic powder, an essential route to prepare dense nanostructured materials by SPS. In: XV International Symposium on Self-Propagating High-Temperature Synthesis. 2019. pp. 124-125. DOI: 10.24411/9999-0014A-2019-10045

[17] Bayrakova O, Golovchenko O, Aknazarov S. Mechano-activation and SHS of difficultly taken ore. Eurasian Chemico-Technological Journal. 2011;**13**(3-4):189-197. DOI: 10.18321/ectj84

[18] Kozhakhmetov Y, Skakov M, Wieleba W, et al. Evolution of intermetallic compounds in Ti-Al-Nb system by the action of mechanoactivation and spark plasma sintering. AIMS Materials Science. 2020;**7**(2):182-191. DOI: 10.3934/matersci.2020.2.182

[19] Tanvar H, Dhawan N. Extraction of rare earth oxides from discarded compact fluorescent lamps. Minerals Engineering. 2019;**135**:95-104. DOI: 10.1016/j.mineng.2019.02.041

[20] Sharma S, Chandra NK, Kumar A, Basu S. Shock-induced atomisation of a liquid metal droplet. Journal of Fluid Mechanics. 2023;**972**:7. DOI: 10.1017/jfm.2023.705

[21] Xiang Y et al. Experimental investigation on debris bed formation from metallic melt coolant interactions. International Journal of Thermal Sciences. 2023;**192**:108398. DOI: 10.1016/j.ijthermalsci.2023.108398

[22] Sakhvadze D, Jandieri G, Tsirekidze T, Gorbenko I. Device for producing metallic powder from melt. GE Patent 20156384B, Oct. 12, 2015

[23] Sakhvadze D, Sakhvadze G, Jandieri G. Method for metallic powder preparation and device for implementation thereof. GE Patent 20207078B, Mar. 10, 2020.

[24] Matsson JE. An Introduction to SOLIDWORKS Flow Simulation. Beginner – Intermediate; 2022. 376 p

[25] Bazhenov I. Automated complex for determination of aluminum activity and gas release kinetics. In: Scientific and Practical Conference "Modern Autoclaved Aerated Concrete" St. Petersburg, 2015, Collection of Reports. pp. 16-19 (in Russ.)

[26] Mironov V, Shishkin A, Polyakov A, Trace Y. Extraction of copper from aqueous solutions using iron powder materials. Ecology. Journal of Belarus. State University. 2018;**1**:97-102 (in Russ.)

[27] Ternovoj J, Bagljuk G, Kudievskij S. Theoretical Basics of Processes of Atomization of Metal Melts: Monografija. Zaporozh'e: Izdatel'stvo ZGIA; 2008. 298 p

[28] Sakhvadze D, Jandieri G, Bolkvadze J. Novel technology of metal powders production by hydrovacuum dispersion of melts. Machines, Technologies, Materials. 2018;**12**(6):236-239. Available from: https://stumejournals.com/journals/mtm/2018/6/236

[29] Sakhvadze D, Jandieri G, Jangveladze G, et al. A new technological approach to the granulation of slag melts of ferrous metallurgy: Obtaining glassy fine-grained granules of improved quality. Journal of Engineering and Applied Science. 2021;**68**:22. DOI: 10.1186/s44147-021-00019-7

[30] Tobo H, Watanabe K, Kuwayama M, Goto S, Goto H, Tanaka T. Effect of water granulation conditions on density and grain size of granulated blast furnace slag. ISIJ International. 2015;**55**(11):2499-2508. DOI: 10.2355/isijinternational.ISIJINT-2015-254

[31] Yu P, Wang S, Li Y, Xu G. A review of granulation process for blast furnace slag. MATEC Web of Conferences. 2016;68:6007. ICIEA. 2016:1-4. DOI: 10.1051/matecconf/20166806007

[32] Pryadko N. Change in material properties at mechanoactivation. Research & Development in Material Science. 2021;**15**(3):RDMS.000862. DOI: 10.31031/RDMS.2021.15.000862

[33] Dang J, Li J, Lv X, Yuan S, Leszczyńska-Sejda K. Metallurgical slag. Crystals. 2022;**12**(3):407. DOI: 10.3390/cryst12030407

[34] Congress of scientists. Together with scientists, we will be able to do the impossible! Information notice. Electronic resource. 2023. Available from: https://civilizebuli.ge/?p=10744 (in Georgian)

Chapter 5

The Physics-Biology Links in Suspended and Settled Estuarine Mud Dynamics: A Review

Eric Wolanski and Michael Elliott

Abstract

In marine, coastal and estuarine environments, it is not possible to understand the ecological processes unless there is a very good understanding of the physical forcing factors. In turn, once the physics of an area creates the conditions for colonisation by the biota, then, very often, the biota modifies the physical and biogeochemical processes. This review details the physical processes involved both in the substratum and the water column, especially in muddy sedimentary systems typical of estuaries and coastal areas. Using recent studies to show that the physical and biological structure and processes work in tandem to create the water column and sedimentary features, the analysis shows that the physics creates the conditions both for chemical mediation and for biological colonisation. The responses on and by the biota include both the micro- and macroorganisms that influence flocculation, floc creation and disintegration, especially through the production of extracellular mucous substances, bioerosion and biostabilisation. Colonisation is firstly by microorganisms and then by macroorganisms. These organisms then create feedback loops where they modify the structure of both the flocs and the bed sediment. It is emphasised that these aspects need to be incorporated into the engineering models of fine sediment dynamics of estuarine and coastal waters, in order to increase their reliability.

Keywords: erosion, settling, flocculation, bioconsolidation, bioerosion, feedback loops, modelling

1. Introduction

Much of the scientific and engineering literature about the physical aspects of cohesive sediment (mud) dynamics comes from laboratory experiments, whereas the biological information comes from laboratory, field and field mescosm studies (e.g., [1]). Clay particles have a diameter <4 μm, whereas silt particles are 4–63 μm, and silt and clay together are usually termed 'mud' [2]. The silt particles have microbial coatings, whereas clay particles have electrostatic charged surfaces, and both types adhere to each other to form flocs in the presence of salt in the water. Such flocs have a diameter of typically 50–1000 μm, and they contain thousands of mud particles. The clay crystallite has positive ions on the face surfaces and negative charges in the interior, but a broken bond at the edges. The formation of the flocs created by clays, in the absence of biology, is due to electric attraction between the dissolved salt ions Na^+ in salt water and the negative charges of the clay particles at the edges [3]. The microbial coating on the silt particles, and to a lesser extent sand grains (>63 µm), creates adhesion between particles which varies with salinity and pH of the water column [4].

Although originally based on the physics indicated by Stoke's Law, more accurately based on laboratory studies of the erosion and settling of mud, engineers have derived formulae for the erosion rate E and the settling rate D [5].

$$E = \left\{0, \text{if } u < u_c A\left(u / u_c - 1\right)^n, \text{if } u > u_c\right\} \tag{1}$$

$$D = \left\{SSC\, w_f\, \left(1u / u_d\right)^2, \text{if } u < u_d; 0, \text{if } u > u_d\right\} \tag{2}$$

In these equations, SSC is the suspended solids concentration, w_f is the settling velocity of the suspended sediment, u is the water speed, u_c and u_d are threshold velocities for respectively entrainment and deposition, A is an erosion parameter (that is generally assumed to have a fixed value for a given mud at a given site) and n is a constant ($2 < n < 4$).

However, when the various parameters and constants in the above equations are derived from laboratory experiments, these equations are inaccurate for field applications [6]. The reason for that is that biological materials and processes and organisms play a dominant role in fine sediment dynamics. This is described and explained here for (a) settling, (b) consolidation, and (c) erosion.

2. The influence of the biological materials in mud dynamics

2.1 Settling

The pioneer laboratory studies [7–9] showed that w_f varies with SSC and that there are two different ranges: firstly, an enhanced settling range whereby for SSC values typically <1 g L^{-1}, where w_f increases with increasing SSC values. Secondly, there is a hindered settling range whereby for larger values of SSC (typically >5 g L^{-1}), w_f decreases with increasing SSC values. The common assumption was that, in the enhanced settling range, the floc size (and hence also the settling velocity if the floc density remains constant) increases with increasing SSC values and that in the longer settling range, the flocs hinder their descent by interfering with each other.

The commercialization of optical backscatter instruments to measure SSC greatly advanced the knowledge of suspended mud particle dynamics in the field as it enabled w_f to be measured from the sequential vertical profile of the suspended particle concentration around slack tide (when the turbulence is the weakest in the tidal cycle); this is derived using a string of optical backscatter profiles or frequent measurements of the vertical profile using a CTD equipped with an optical backscatter. These field studies indicated that the settling velocity derived from laboratory studies differs by up to an order of magnitude from that measured in the field. The enhanced settling range is shown in **Figure 1**.

The difference by an order of magnitude of the floc settling velocity in nature and in laboratory experiments is the result of the influence of biological interactions. Firstly, biology-generated flocculation is very rapid (**Figure 2**). Secondly, biology-derived flocculation increases the size of the flocs by orders of magnitude from that due to salt-induced flocculation and is determined not by fractal properties (such as prevails for snowflakes)

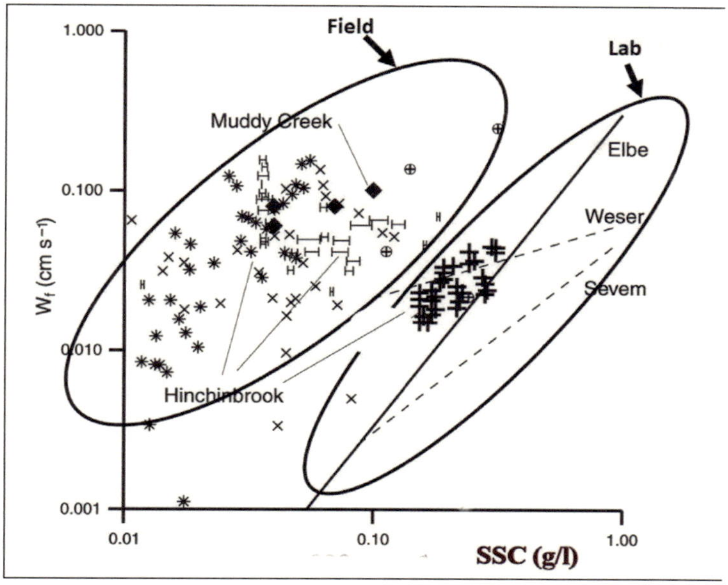

Figure 1.
Dependence of the settling velocity w_f of the suspended sediment on the suspended sediment concentration (SSC) in the enhanced settling range from (1) laboratory experiments with mud from the Elbe, Weser and Severn estuaries and (+) from Hinchinbrook Channel and (2) in the field in Hinchinbrook Channel and Muddy Creek, tropical Australia. The value of w_f in the laboratory is comparable for all the sampled estuaries, and it is about an order of magnitude smaller than that in the field. Modified from [10].

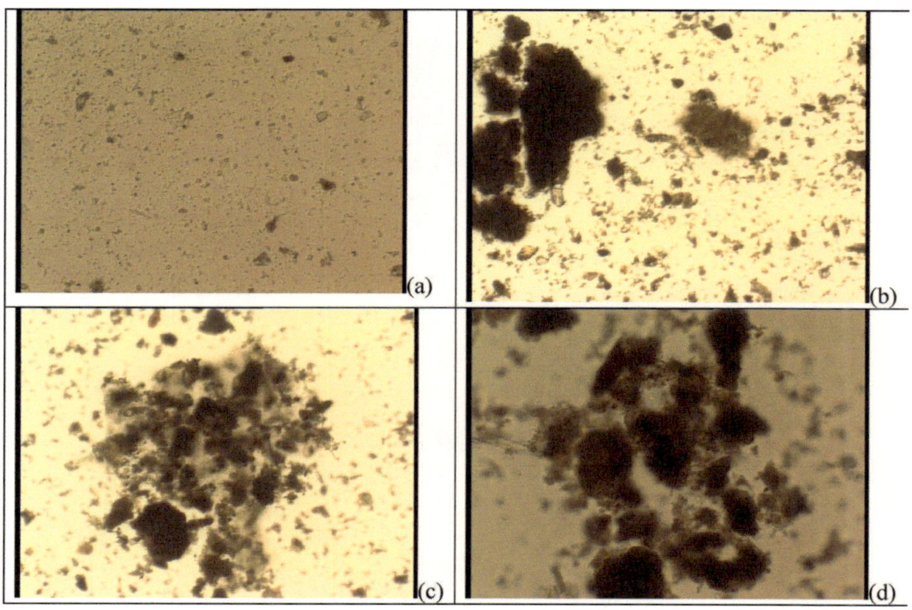

Figure 2.
Microphotographs of suspended fine sediment in the muddy La Sa Fua River in (a) freshwater about to enter Fouha Bay in Guam, (b) mixed with 17 ppt salinity Fouha Bay water after (b) 5 min, (c) 10 min and (d) 30 min. The width of the microphotographs is 1000 μm.

but instead by the biological materials on the surface and inside the flocs (**Figure 3**). As shown in **Figure 3**, the mucus has produced a complex shape and size of the flocs.

Each floc is a matrix (mesh-like structure) of inorganic and organic matter, forming a maze of low-density zones and impervious zones. The settling floc displaces water, some of which flows around the floc and some through the floc [1]. The extreme situation of biological control of the settling occurs for large sticky mucus sheets with a few attached particles of mud that act as a ballast. This makes the mucus sheet form an inverse parachute dragged downward by the settling floc [11].

In nature, as opposed to laboratory experiments, the estuarine flocs appear as marine snow flocs of fine (silt and clay) particles. The particles are bound together with a sticky mucus membrane called EPS and TEP (extra-cellular polymeric substances and transparent exopolymer particles from bacteria and diatoms). Hence,

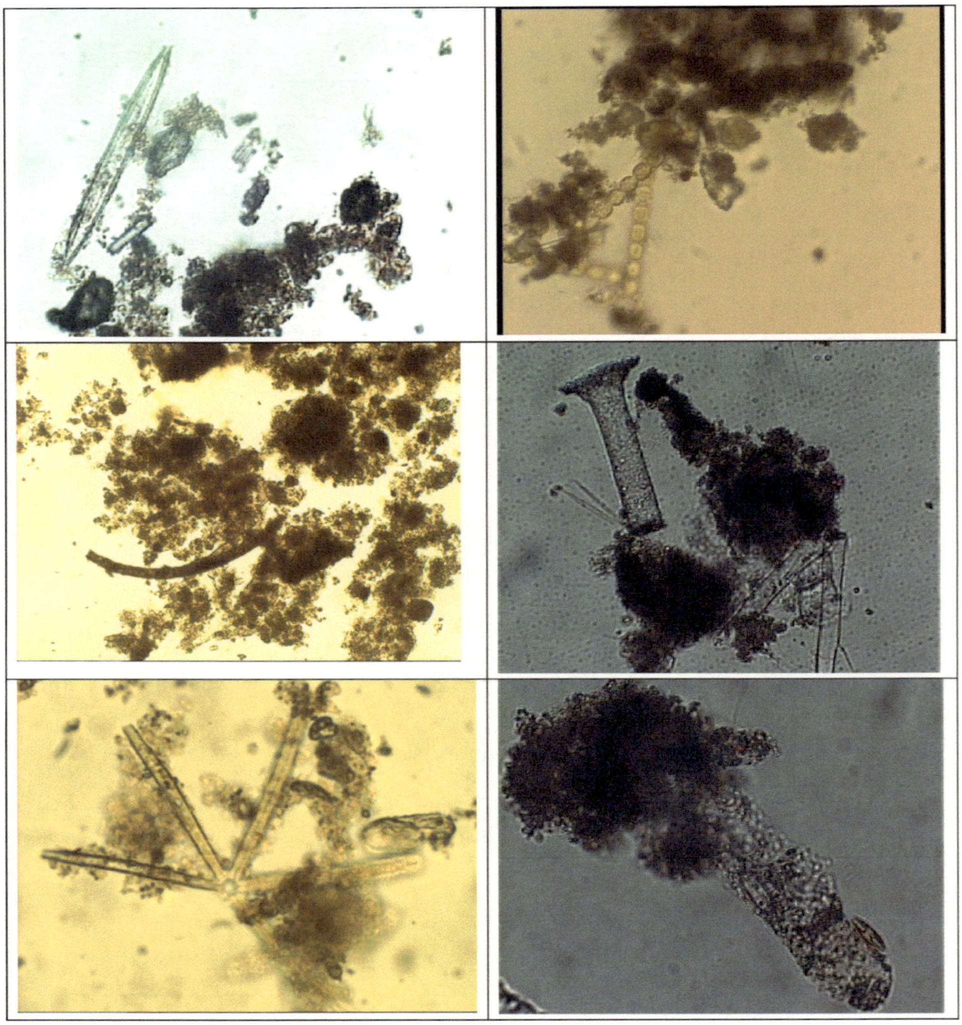

Figure 3.
Microphotographs of typical flocs in muddy estuaries and coastal waters in Papua New Guinea and tropical Australia. The width of the microphotographs varies between 500 and 1000 μm.

the floc becomes a complex mixture of microbial coatings, absorbed and adsorbed organic matter, silt and clay particles, dead and live plankton and macroscopic aggregates of biological origin [1, 12]. The floc structure is fundamentally controlled by the mucus (mucopolysaccharides/glycosaminoglycans) [13].

The formation of larger and faster sinking flocs is facilitated by the fresh TEP, which is highly sticky and traps small flocs that small-scale turbulence brings in contact. In a temperate climate, the smallest amount of fresh TEP and the smallest floc size occur in winter when the biological activity is the smallest [14]. Small mud flocs grow into large muddy marine snow flocs by using the mucus to stick on the clay and silt particles and the dead organic matter [15–21]. In some cases, clay and silt particles are segregated in small sub-units held together by mucus. Fine sediment flocs can reach as large as 2 cm in length when they aggregate over dead chain-forming diatoms [22]. The porosity of such flocs is high. In turbid waters (SSC > 500 mg l^{-1}), the flocs are smaller, because of the lack of light for photosynthesis.

As the small flocs cluster together with increasing turbulence, the growing aggregate is more porous and its strength reduced. Consequently, these large but weak aggregates are broken by a high shear rate once the turbulent kinetic energy dissipation rate exceeds a threshold value of the order of 10^{-6} W kg^{-1} [23, 24]. Thus, the floc size varies within a tidal cycle because the turbulence also varies within that cycle [25]. At the same time, the biological processes, materials and organisms constantly change the flocs in time [1]. The influence of organisms is dictated by their biological traits (their mobility, mode of feeding and position in the water column). For example, the zooplankton feed on the flocs and on the mucus (**Figure 4a** and **b**); also, they move through the floc while the mud sticks on their bodies, thus changing the shape of the flocs (**Figure 4c**); also, faecal pellets produced by the zooplankton are created by aggregating small flocs (**Figure 4d**); the plankton live in the high-mucus, low-density/porous area of the muddy marine snow floc that it disturbs (**Figure 4e**), and the plankton change the shape and density of the floc by being 'glued' to it (**Figure 4f**). The biological materials are a fundamental building block of muddy marine snow. These biological processes then influence the size and shape of the flocs, thereby affecting the way in which the flocs are influenced by the physical processes, a de facto feedback loop. In essence, the physical and microbiological processes create the flocs and then the macrobiology (the organism behaviour) modifies the flocs which in turn creates a further change by physical processes. This gives a physical-biological-physical loop. Secondly, as indicated above, are the interactions with the physical processes – the turbulence affects the floc size because, at small turbulence intensity, it helps small flocs collide with each other and aggregate to form larger flocs, and, at high levels of turbulence (e.g., during peak tidal currents or under wave action), it breaks up the larger flocs into smaller flocs [25].

The microbial loop is a central process in using and recycling nutrients in the water column [27], and its efficiency relies on the amount of surface available over which it occurs; this is particularly important in the turbidity maximum areas of estuaries [28]. The flocs may be formed from organic detritus as predominantly carbon, but with increasing bacterial colonisation, the nitrogen component increases, that is, with age, there is often a decrease in the C/N ratio. In turn, this increases the nutritional value of the flocs for planktonic and macrobenthic suspension feeding organisms. Given the increasing surface area:volume ratio with decreasing size of particle, smaller particles have a large surface area on which the bacterial film can form.

The density of the organic-rich flocs varies from floc to floc [21, 29, 30]. For instance, faecal pellets are formed by suspension-feeding benthic bivalves and are much denser and have a faster settling velocity than other biology-rich flocs, whereas the looser

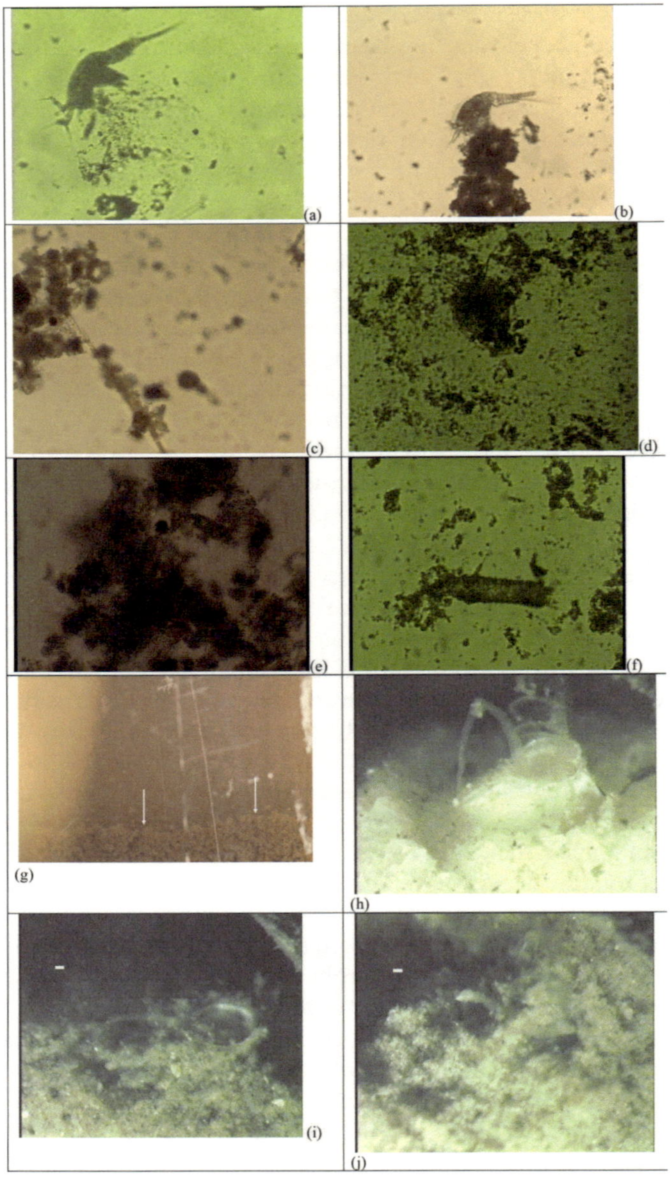

Figure 4.
Video clips available at https://intech-files.s3.amazonaws.com/a043Y00000yGSwYQAW/a093Y00001h6OeqQAE/Videos%20%28Full%20Chapter%29-The%20physics-biology%20links%20in%20fine%20sediment%20dynamic%20%282024-04-23%2015%3A48%3A53%29.zip. (a) The mucus enveloping small sediment particles attracts a copepod that feeds on it. (b) A copepod feeding on a muddy marine snow floc. (c) A plankton moving through a mud floc with mud sticking on its body and preventing it from escaping. (d) The process of pelletisation. (e) A plankton immersed in a highly porous, muddy marine snow floc comprising clay, silt particles and TEP. (f) A plankton stuck in a sticky floc and making it in increase in size by more mud sticking on it. (g) Consolidation processes of the highly organic Lake Apopka mud, Florida. The view shows the sediment both above and below the bed. Flocs above the bed are settling down. The left arrow points to a growing micro-channel. Water moves upwards in the micro-channel and rolls flocs upwards along the walls of the micro-channel to keep it open. The right arrow points to a channel that is getting blocked by flocs settling in its opening and plugging it. (h–j) settling muddy marine snow flocs stick to the tentacle-like cirri of a Pyrgomatid barnacle and smother it as the barnacle is unable to dislodge the sticky flocs and sedimentation continues. (a) and (b) are adapted from [26].

pseudo-faeces produced are mucus-rich and they add to the floc material. Further, their rate of production is not constant; for SPM < 300 mg l^{-1}, the filtration rate by bivalves increases with increasing SPM, but it decreases for SPM > 300 mg l^{-1}. Also, macro-flocs aggregated by the mucus (EPS and TEP) appear to have the lowest density, then likely the smallest settling velocity. Further, zooplankton break up flocs by feeding on them and thus decrease the floc settling velocity, while other plankton break flocs by moving through them while being glued to them by the mucus (**Figure 4a–f**).

While physical and physico-chemical processes can create the sedimentary systems, both on the bed and in the water column, it is then that biological processes modify those physical processes with feedback loops [1]. Previously, engineers neglected these interactions—for example, they applied the fractal theory to mud particle flocs to parameterise their shape and size, assuming thus that mud flocs are similar to snowflakes and that the biology can be ignored (e.g., [31]). However, from **Figures 2, 3** and **4a–f**, it is clear that there is little in natural estuaries and coastal waters that visually resembles the 'fractal flocs' derived from theoretical or laboratory models with no biology. That the biology has a dominant effect on the settling velocity w_f is evidenced by the failure of models to predict w_f from flocculation models and fractal theory that ignore the biology. The common observation from field studies is that the floc size varies during the tidal cycle [9, 32].

Clearly, w_f cannot be measured in the laboratory and from models where the biology is neglected. There is a difference by a factor of 5–10 between the observed settling velocity of mud flocs with and without the biology (**Figure 1**). For all these biological and physical reasons, as indicated above, the settling velocity of individual mud flocs at a given site also varies greatly during a tidal cycle [33]. In the hindered settling range, the settling of flocs strongly depends on the density, hence on the biology as biological-rich flocs have a much lower density than inorganic flocs. At SSC > 5 g L^{-1}, the settling flocs align themselves one behind the other, each floc settling in the lee of the preceding floc (**Figure 5**; [34]). Water moves upwards in micro-channels. In the presence of turbulence, settling is locally stopped by the turbulence moving a floc sideways in the micro-channel and blocking it.

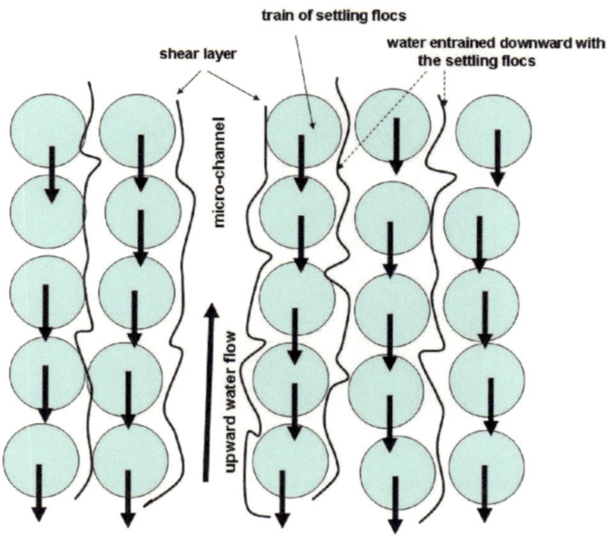

Figure 5.
Settling of high-density mud flocs for SSC > 5 g L^{-1}. Adapted from [1].

2.2 Consolidation

The settled mud is initially water-rich as the flocs are very porous. It is thus readily erodible. It slowly consolidates by expelling water through micro-channels. This process is very slow; typically, it takes 3 weeks to 3 months per m for organic-poor mud. For organic-rich mud, it can take up to 20 years [35] because inorganic mud, i.e., the mineral particles, is heavier and therefore it expels water faster sideway towards the micro-channels than organic mud does. Dewatering occurs when the water moves upwards through micro-channels (**Figure 6**). While doing so, it rolls mud flocs upwards along the walls of the micro-channel to finally expel them, thus widening the micro-channels and accelerating the dewatering (**Figure 4g**). Occasionally, a micro-channel can get plugged on top by mud flocs in the overlying water column settling in the opening of the micro-channel (**Figure 4g**).

Figure 6.
Photograph of a dewatering micro-channel in Lake Apopka mud. The mud is consolidating by expelling water sideway from the flocs, and this water flows upwards in the micro-channel.

Mud deposition at tidal frequencies commonly forms laminae several cm thick, therefore creating strata (laminae) (**Figure 7**). Each lamina was deposited in a discrete hydrodynamic event. Each lamina has a different density varying between 1.4 and 2.1 g cm^{-3} [36]. Older (deeper) laminae can have a smaller density than overlying laminae, because a new lamina settled on top of another one can block its dewatering micro-channels and hence prevent consolidation.

Figure 7.
(a) Photograph of muddy laminae on the banks of the Ord estuary, Australia. (b) Vertical profile of density as a function of depth through a sequence of muddy laminae in the Scheldt estuary, Belgium-the Netherlands, adapted from [36].

2.3 Erosion

Once the particles have settled, the sediment created is subject to physical, chemical and biological forces and interactions (**Figure 8**). Microorganisms continue to coat the particles and lead to biogeochemical processes, while macroorganisms will colonise the sediments according to the environmental tolerances and biological traits of the organisms [37, 38]. Biological processes by those macroorganisms such as bioturbation, biodeposition and bioerosion will dictate the nature of the sediments; bioturbation (the process of burrowing and sediment turnover by organisms) will aerate cohesive sediments and therefore change the chemical regime. Cohesive sediments with a low porosity and poor permeability will become anoxic, as shown by the position of the redox potential discontinuity [1, 12]. The erosion rate is greatly affected by the biological activity in the bottom mud. The bed can be destabilised or stabilised, according to the specific biological activity although wider climatic and ambient conditions will also influence these characteristics [12, 39, 40].

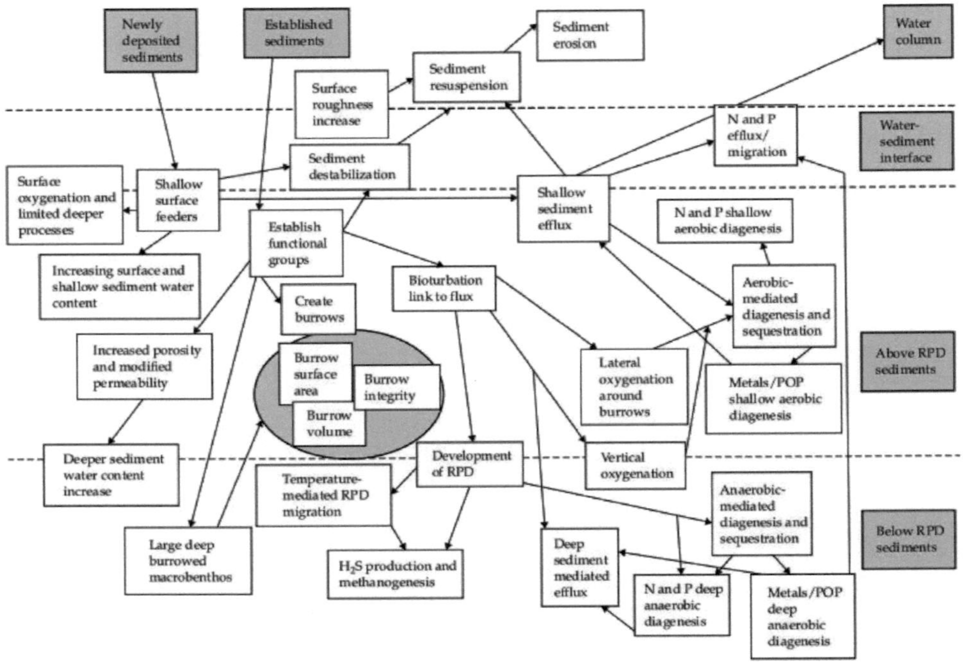

Figure 8.
Physical, chemical and biological processes in cohesive sediments. Adapted from [12].

2.3.1 Destabilisation by bioturbation

At a microorganism level, diatoms produce oxygen that can form an air bubble in the sediment. The bubble loosens and even dislodge the sediment. This increases the erosion rate [41]. At a microorganism level, burrowing animals disturb the sediment to a depth of typically ~1 m for mangrove crabs, down to ~ a few mm for meiofauna; erosion is increased by the feeding behaviours of large bivalves (**Figure 9**; [39, 40, 43–45]). The sediment is deposited at the surface from where it is easily eroded [21]. The small bioturbating macrofauna are apparently the most efficient organisms to erode a highly muddy substratum [46]. Biological processes may initiate the formation of tidal creeks in the case of the crab *Chasmagnathus granulatus* digging burrows. A depression ring results, which grows and links with other rings; in turn, this directs the tidal flows that excavate a tidal creek [47, 48].

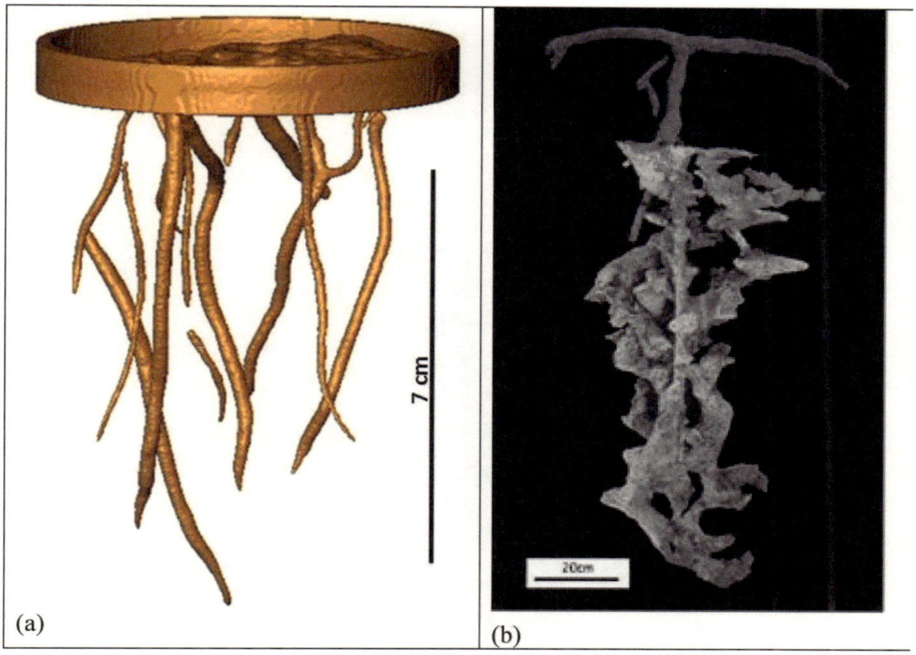

Figure 9.
(a) A 10 cm deep burrow structure produced by H. diversicolor as revealed by computer-aided tomography, adapted from [42] (b) A 1.2 m deep mangrove mud crab burrow. Adapted from [1].

2.3.2 Stabilisation by filtering the mud and armouring the bed

Conversely, the biological processes and organisms can stabilise the mud [21]. Algae mats and biofilms armour the surface of mud banks and protect the sediment from erosion (**Figure 10a**; [30, 49, 50]). Benthic diatoms biostabilise the mud surface [51, 52].

Especially as the result of mucopolysaccharide production. However, this process is reversed when the amphipod *Corophium volutar* feeds on the benthic diatoms that bind the sediment particles; in turn, this process can be controlled by birds feeding on this amphipod. Tube worms such as Terebellids or Sabellids whose tube protrude above the sediment surface also affect the boundary layer currents and may reduce erosion [12]. Fauna living on the bottom also trap the mud flocs because they are sticky. For instance, barnacles trap the mud flocs as they stick on their tentacle-like

Figure 10.
(a) A biofilm on a tropical mud flat armouring the underlying mud flat. Source: Norman Duke. Adapted from [1]. (b) A mussel bed armouring the mud. Source: Wadden Sea National Park of Lower Saxony. (c) The leather mudweed Avrainvillea amadelpha trapping mud in Maunalua Bay, Hawaii. (d) The dense vegetation of a Rhizophora mangrove forest. (e) The dense salt marsh vegetation along a tidal creek. Source: Steve Hillebrand.

cirri that they try to dislodge, a process that is increasingly less successful with increased sedimentation rate (**Figure 4h–j**; [19]). In addition to producing faecal pellets and pseudofaeces, oyster and mussel beds armour the bed, thus preventing erosion (**Figure 10b**; [53–55]). The leather mudweed *Avrainvillea amadelpha,* a marine invader in Maunalua Bay, Hawaii, traps 0.2 kg of mud /m^2 below the canopy for years and another 0.2 kg of mud/m^2 in the canopy (**Figure 10c**; [56]).

The mangrove or saltmarsh vegetation fringing the open water of an estuary is a mud trap due to the small-scale recirculating flows behind the vegetation that create flow stagnation, which in turn promotes the settling of mud (**Figure 10d**-ed-e; [1, 57–63]). The trapping of mud in this dense vegetation is most efficient; for instance, a mangrove forest that covers only 3.8% of the river drainage area traps 40% of the riverine mud inflow [63, 64]. Mangroves can trap up to 1000 tonnes of mud km^{-2} yr^{-1} [63, 64]. Similar processes occur in saltmarsh-fringed estuaries, and especially so when the saltmarsh vegetation is taller than 8 cm [65–69]. The rates of sedimentation in U.K. saltmarshes average 4.3 mm yr^{-1} over two years of observation, with large local spatial and temporal fluctuations [70, 71]. A similar high sedimentation rate is found in U.S. saltmarshes [72].

Also, at the macro-scale, biology also helps mitigate bank erosion, for which no reliable engineering formulae are available. Mass slippage often occurs at low water after a river flood when the soil is saturated and its weight is increased by the groundwater [73]. In macro- and hyper-tidal estuaries, the banks experience a frequent transition between stability at high tide and instability at low tide, and this also leads to bank slippage [74]. Nevertheless, some estuaries with steep banks remain stable against bank slippage because slippage is hindered by the roots of the trees and plants on the shores that increase the cohesiveness of the soils, by the vegetation covering

Figure 11.
An aerial photograph of the dying freshwater vegetation and the resulting bare mud areas following the permanent intrusion of a tidal creek in the formerly freshwater floodplains of the Mary River, Northern Territory, Australia.

the banks and protecting them from the swift currents, and by the organics from decaying roots making the soil more cohesive [1, 75–79]. When the freshwater region of an estuary transitions permanently from fresh to saline, such as in the Mary River, Northern Territory, Australia, its freshwater vegetation is killed, thus removing the protection that the freshwater vegetation provided. This results in swift bank erosion, in the lengthening of the tidal creek in the formerly freshwater floodplains and in the formation of new, bare mud flats that erode and provide new mud to the estuary (**Figure 11**; [80]). However, in many estuaries, the freshwater vegetation will give way to reedbeds, such as *Phragmites australis*, and then saltmarsh and seagrass beds with a progression down the estuary and increasing salinity. While reedbeds may create 1.5 high reeds during the growing season, at the end of the growing season, the reeds collapse and are added to the detrital pool of the estuary. The stubble (broken straws) left behind then traps sediment, reduces bed erosion and stabilises the reed-bed area [81].

3. Some engineering implications

The long-term sequestration in the mud of persistent organic pollutants, radionuclides and heavy metals on the fine sediment is a serious problem for industrialised countries attempting to clean up their estuaries from legacy pollution that started with industrialisation 100–200 years ago [82, 83]. Those contaminants may be stored in the sediments through diagenetic processes, especially if these sediments are anoxic and consolidated, thereby creating biogeochemical conditions leading to insoluble stored products [27]. Aeration by bioturbation will then change the oxygenated regime and so liberate newly soluble products. Hence, bioturbation is important in slowly cleaning up this polluted mud by allowing ventilation of the mud through animal burrows and increasing the depth of oxidation of the mud [84]. The flushing out of these pollutants depends on the biology through the occurrence of anoxia generated by plankton or algae blooms, the role of bacteria, DOM and the presence of macroalgae and macrophyte roots, all of which generally decrease metal mobility [85–88]. However, not all these natural processes are beneficial; for instance, the bacterial communities convert mercury to its more toxic methylmercury form [89] and methanogenic and reducing bacteria will produce methane CH_4 and the more toxic hydrogen sulphide H_2S. Also, some of these processes will prevent permeability and hence oxygenation of the sediments, and so the sediments remain anoxic.

Cholera is still causing morbidity and mortality in many developing countries. Its causative agents are *Vibrio cholerae* O1 and O139. It is transmitted mainly through water and by eating shrimps and fish that host the bacteria. The zooplankton responding to anthropogenic inputs controls cholera incidence [90]. In addition, the resuspension of fine sediments from polluted tidal flats raises *Vibrio cholera* concentration in the water, which suggests a *Vibrio cholera*-coupled mud-pelagic dynamics [91].

Fluid mud impacts the propagation of tides in an estuary by diminishing bottom friction [92–94]. Its removal by dredging can make the estuary self-deepening, which in turn generates severe environmental and socio-economic problems [95].

4. Concluding remarks

It is emphasised here that in all estuarine and coastal sedimentary systems, the physical and biological structure and processes work in tandem to create the resulting water column and sedimentary features. The physics then creates the conditions both for chemical mediation and for biological colonisation that occurs in all estuarine and coastal environments worldwide and is dependent on the prevailing temperature, salinity and pH conditions. That colonisation will be firstly by microorganisms and then by macroorganisms. These organisms then always create feedback loops where they modify the structure of both the flocs and the bed sediment. This changes the behaviour of the flocs and the bed sediment features. This should always be incorporated into the engineering models of fine sediment dynamics of estuarine and coastal waters, for those models to become reliable.

Land use in the catchment upstream of estuaries and along the shores of estuaries, land claim in coastal wetlands, pollution, dams/barriers and dredging, among many other causes, have profoundly degraded many estuaries worldwide [96]. There are attempts worldwide in many, if not all, maritime states to restore the ecosystem health of the degraded estuaries [97, 98]. The degradation of estuaries in the developed world started in the 1700s with land claim for agriculture and then continued through the Industrial Revolution of the 1800s with the addition of contaminants from industries. Many of the underlying causes for degradation are due to mishandling the mud either at the source (e.g., erosion) or at the sink (deposition). Most commonly, the experience suggests that the solution involves combining the use of plants and animals and acknowledging the role of the microbial populations and the materials that they exude, together with wise engineering, eco-engineering (i.e. nature-based solutions) and geo-engineering measures [1, 99–103].

Author details

Eric Wolanski[1]* and Michael Elliott[2,3]

1 College of Science and Engineering, James Cook University, Townsville, Queensland, Australia

2 School of Environmental Sciences, University of Hull, UK

3 International Estuarine and Coastal Specialists (IECS) Ltd, Leven, UK

*Address all correspondence to: eric.wolanski@jcu.edu.au

References

[1] Wolanski E, Elliott M. Estuarine Ecohydrology: An Introduction. Amsterdam: Elsevier; 2015. p. 322

[2] Uncles RJ, Mitchell SB, editors. Estuarine and Coastal Hydrography and Sediment Transport. Cambridge: Cambridge University Press; 2017. p. 351

[3] van Olphen H. Internal mutual flocculation in clay suspensions. Journal of Colloid Science. 1964;**19**(4):313-322

[4] Whitehouse R, Soulsby R, Roberts W, Mitchener H. Dynamics of Estuarine Muds. London: Thomas Telford; 2000. p. 210

[5] Partheniades E. Erosion and deposition of cohesive soils. Journal of the Hydraulics Division, American Society of Civil Engineers. 1965;**91**:105-138

[6] Wolanski E. Estuarine ecohydrology modeling: What works and within what limits? In: Wolanski E, Day JW, Elliott M, Ramachandran R, editors. Coasts and Estuaries: The Future. Amsterdam: Elsevier; 2019. pp. 503-521

[7] Mehta AJ. On estuarine cohesive sediment suspension behavior. Journal of Geophysical Research Ocean. 1989;**94**(C10):14303-14314

[8] Winterwerp JC, Bale AJ, Christie MC, Dyer KR, Jones S, Lintern DG, et al. Flocculation and settling velocity of fine sediment. In: Winterwerp JC, Kranenburg C, editors. Fine Sediment Dynamics in the Marine Environment. Proceedings in Marine Science 5. Amsterdam: Elsevier; 2002. pp. 25-40

[9] Winterwerp JC. The physical analyses of muddy sedimentation processes. In: Uncles RJ, Monismith SG, editors. Water and Fine Sediment Circulation. Treatise on Estuarine and Coastal Science 2. Amsterdam: Elsevier; 2011. pp. 311-350

[10] Syvitski JPM, Harvey N, Wolanski E, Burnett WC, Perillo GME, Gornitz V. Dynamics of the coastal zone. In: Crossland, CJ, Kremer HH, Lindeboom HJ, Marshall Crossland JI, Le Tissier MDA, Coastal Fluxes in the Anthropocene. Berlin: Springer-Verlag; 2005. p. 39-94

[11] Asaeda T, Wolanski E. Sinking of muddy marine snow. Wetlands Ecology and Management. 2002;**10**:283-287

[12] Gray JS, Elliott M. Ecology of Marine Sediments: From Science to Management. Oxford: Oxford University Press; 2009. p. 256

[13] Ye L, Penaloza-Giraldo JA, Manning AJ, Holyoke J, Hsu TJ. Biophysical flocculation reduces variability of cohesive sediment settling velocity. Communications Earth & Environment. 2023;**4**:138

[14] Fettweis M, Schartau M, Desmit X, Lee BJ, Terseleer N, Van der Zande D, et al. Organic matter composition of biomineral flocs and its influence on suspended particulate matter dynamics along a nearshore to offshore transect. Journal of Geophysical Research: Biogeosciences. 2022;**127**:e2021JG006332

[15] Ayukai T, Wolanski E. Importance of biologically mediated removal of fine sediments from the Fly River plume, Papua New Guinea. Estuarine, Coastal and Shelf Science. 1997;**44**:629-639

[16] Wolanski E, Spagnol S, Ayukai T. Field and model studies of the fate

of particulate carbon in mangrove-fringed Hinchinbrook Channel, Australia. Mangroves and Salt Marshes. 1998;**2**:205-221

[17] Wolanski E, Richmond R, McCook L, Sweatman H. Mud, marine snow and coral reefs. American Scientist. 2003;**91**:44-51

[18] Wolanski E, Richmond RH, Davis G, Bonito V. Water and fine sediment dynamics in transient river plumes in a small, reef-fringed bay, Guam. Estuarine, Coastal and Shelf Science. 2003;**56**:1029-1043

[19] Fabricius K, Wolanski E. Rapid smothering of coral reef organisms by muddy marine snow. Estuarine, Coastal and Shelf Science. 2000;**50**:115-120

[20] Pejrup M, Mikkelsen OA. Factors controlling the field settling velocity of cohesive sediments in estuaries. Estuarine, Costal and Shelf Science. 2010;**87**:177-185

[21] Andersen TJ, Pejrup M. Biological influences on sediment behavior and transport. In: Uncles RJ, Monismith SG, editors. Water and Fine Sediment Circulation. Treatise on Estuarine and Coastal Science. Vol. 2. Amsterdam: Elsevier; 2011. pp. 289-309

[22] Berhane I, Sternberg RW, Kineke GC, Milligan TG, Kranck K. The variability of suspended aggregates on the Amazon continental shelf. Continental Shelf Research. 1997;**17**:267-285

[23] Tansel B. Morphology, composition and aggregation mechanisms of soft bioflocs in marine snow and activated sludge: A comparative review. Journal of Environmental Management. 2018;**205**:231e243

[24] Takeuchi M, Doubell MJ, Jackson GA, Yukawa H, Sagara Y, Yamazaki H. Turbulence mediates marine aggregate formation and destruction in the upper ocean. Scientific Reports. 2019;**9**:16280

[25] Winterwerp JC. On the flocculation and settling velocity of estuarine mud. Continental Shelf Research. 2002;**22**(9):1339-1360

[26] Wolanski E, Kingsford M, Lambrechts J, Marmorino G. The physical oceanography of the great barrier reef: A review. In: Wolanski E, Kingsford MJ, editors. Oceanographic Processes of Coral Reefs. Physical and Biological Links in the Great Barrier Reef. Boca Raton: CRC Press; 2024. pp. 9-34

[27] Libes SM. An Introduction to Marine Biogeochemistry. 2nd ed. London: Academic Press; 2009. p. 909

[28] McLusky DS, Elliott M. The Estuarine Ecosystem: Ecology, Threats and Management. 3rd ed. Oxford: Oxford University Press; 2004. p. 214

[29] Andersen TJ. Seasonal variation in erodibility of two temperate, microtidal mudflats. Estuarine, Coastal and Shelf Science. 2001;**53**:1-12

[30] Andersen TJ, Lund-Hansen L, Pejrup M, Jensen KT, Mouritzen KN. Biologically induced differences in erodibility and aggregation of subtidal and intertidal sediments: A possible cause for seasonal changes in sediment deposition. Journal of Marine Systems. 2005;**55**:123-138

[31] Zhu Z. A formula for the settling velocity of cohesive sediment flocs in water. Water Supply. 2019;**19**(5):1422-1428

[32] Mikes D, Manning AJ. Assessment of flocculation kinetics of cohesive sediments from the seine and Gironde

estuaries, France, through laboratory and field studies. Journal of Waterway, Port, Coastal, and Ocean Engineering. 2010;**136**(6):306-318

[33] Manning AJ, Martens C, De Mulder T, Vanlded J, Wintewerp J, Ganderton P, et al. Mud floc observations in the turbidity maximum zone of the Scheldt estuary during neap tides. Journal of Coastal Research. 2007;**SI50**:832-836

[34] Wolanski E, Gibbs R, Mazda Y, Mehta A, King B. The role of buoyancy and turbulence in the settling of mud flocs. Journal of Coastal Research. 1992;**8**:35-46

[35] Mehta AJ, Jaeger JM, Valle-Levinson A, Hayter EJ, Wolanski E, Manning AJ. Resuspension Dynamics in Lake Apopka, Florida. Report No. UFL/COEL-2009/001, Department of Civil and Coastal Engineering. Gainesville: University of Florida; 2009. p. 59

[36] Wartel S. Mud Layer and Cyclic Sedimentation Patterns in the Estuary of the Schelde. Belgium – The Netherlands; 2003. Available from: http://www.naturalsciences.be/sedimento/publications/Abstract_1

[37] Widdows J, Brinsley MD, Salkeld PN, Elliott M. Use of annular flumes to determine the influence of current velocity and bivalves on material fluxes at the sediment-water interface. Estuaries. 1998;**21**(4A):552-559

[38] Widdows J, Brinsley M, Elliott M. Use of in situ flume to quantify particle flux (biodeposition rates and sediment erosion) for an intertidal mudflat in relation to changes in current velocity and benthic macrofauna. In: Black KS, Patterson DM, Cramp A, editors. Sedimentary Processes in the Intertidal Zone. Special Publications 139. London: Geological Society; 1998. pp. 85-97

[39] Widdows J, Brown S, Brinsley MD, Salkeld PN, Elliott M. Temporal changes in intertidal sediment erodability: Influence of biological and climatic factors. Continental Shelf Research. 2000;**20**:1275-1289

[40] Widdows J, Brinsley MD, Salkeld PN, Lucas CH. Influence of biota on spatial and temporal variation in erodibility and material flux on a tidal flat (Westerschelde, the Netherlands). Marine Ecology Progress Series. 2000;**194**:23-37

[41] Perillo GME. Tidal courses: Classification, origin and functionality. In: Perillo GME, Wolanski E, Cahoon DR, Brinson MM, editors. Coastal Wetlands: An Integrated Ecosystem Approach. Amsterdam: Elsevier; 2009. pp. 185-210

[42] Mazik K, Curtis N, Fagan MJ, Taft S, Elliott M. Accurate quantification of the influence of benthic macro- and meio-fauna on the geometric properties of estuarine muds by micro computer tomography. Journal of Experimental Marine Biology and Ecology. 2008;**354**:192-201

[43] Orvain F. A model of sediment transport under the influence of surface bioturbation: Generalisation to the facultative suspension-feeder *Scrobicularia plana*. Marine Ecology Progress Series. 2005;**286**:43-56

[44] Ciutat A, Widdows J, Readman JW. Influence of cockle *Cerastoderma edule* bioturbation and tidal-current cycles on resuspension of sediment and polycyclic aromatic hydrocarbons. Marine Ecology Progress Series. 2006;**328**:51-64

[45] Soares C, Sobral P. Bioturbation and erodibility of sediments from the Tagus estuary. Journal of Coastal Research. 2009;**SI**(56):2429-1433

[46] Harris J, Pilditch CA, Greenfield BL, Moon V, Kröncke I. The influence of benthic macrofauna on the erodibility of intertidal sediments with varying mud content in three New Zealand estuaries. Estuaries and Coasts. 2016;**39**:815-828

[47] Perillo G, Minkoff DR, Piccolo MC. Novel mechanism of stream formation in coastal wetlands by crab–fish–groundwater interaction. Geo-Marine Letters. 2005;**25**:214-220

[48] Minkoff DR, Escapa MJ, Ferramola FE, Maraschín SD, Pierini JO, Perillo GME, et al. Effects of crab–halophytic plant interactions on creek growth in a S.W. Atlantic salt marsh: A cellular automata model. Estuarine, Coastal and Shelf Science. 2006;**69**:403-413

[49] Lumborg U, Andersen TJ, Pejrup M. The effect of *Hydrobia ulvae* and microphytobenthos on cohesive sediment dynamics on an intertidal mudflat described by means of numerical modeling. Estuarine, Coastal and Shelf Science. 2006;**68**:208-220

[50] Venier C, Figueiredo da Silva H, McLelland SJ, Duck RW, Lanzoni S. Experimental investigation of the impact of macroalgal mats on flow dynamics and sediment stability in shallow tidal areas. Estuarine, Coastal and Shelf Science. 2012;**112**:52-60

[51] Blanchard GF, Paterson DM, Stal LJ, Richard P, Galois R, Huet V, et al. The effect of geomorphological structures on potential biostabilisation by microphytobenthos on intertidal mudflats. Continental Shelf Research. 2000;**20**:1243-1256

[52] Christie MC, Dyer KR, Blanchard G, Cramp A, Mitchener HJ, Paterson DM. Temporal and spatial distributions of moisture and organic contents across a macro-tidal mudflat. Continental Shelf Research. 2000;**20**:1219-1241

[53] van Leeuwen B. Mussel Bed Influence on Fine Sediment Dynamics on a Wadden Sea Intertidal Flat. Delft: Deltares and University of Twente; 2008. p. 110

[54] Kangeri AK, Jansen JM, Joppe DJ, Dankers ANMJA. In situ investigation of the effects of current velocity on sedimentary mussel bed stability. Journal of Experimental Marine Biology and Ecology. 2016;**485**:65-72

[55] Commito JA, Dankers NMJA. Dynamics of spatial and temporal complexity in European and north American soft-bottom mussel beds. In: Reise K, editor. Ecological Comparisons of Sedimentary Shores. Ecological Studies. Vol. 2001. Berlin: Springer-Verlag; 2001. pp. 39-59

[56] Macduff S, Wolanski E, Richmond RH. Invasive alga removal accelerates sediment flushing in Maunalua Bay, Hawaii. TOS/ASLO/AGU Open Sciences Meeting. 2012:B0821. Available from: https://igert2012.videohall.com/presentations/308

[57] Pethick JS. Saltmarsh geomorphology. In: Allen JRL, Pye K, editors. Saltmarshes: Morphodynamics, Conservation and Engineering Significance. Cambridge: Cambridge University Press; 1992. pp. 41-62

[58] Pethick J, Leggett D, Husain L. Boundary layers under salt marsh vegetation developed in tidal currents. In: Thorne J, editor. Vegetation and Erosion Processes and Environments. London: John Wiley; 1992. pp. 113-124

[59] Bridges PH, Leeder MR. Sedimentary model for intertidal mudflat channels with examples from the Solway Firth, Scotland. Sedimentology. 1977;**23**:533-552

[60] Furukawa K, Wolanski E, Muller H. Currents and sediment transport in mangrove forests. Estuarine, Coastal and Shelf Science. 1997;**44**:301-310

[61] Fagherazzi S, Wiberg PL, Temmerman S, Struyf E, Zhao Y, Raymond PA. Fluxes of water, sediments, and biogeochemical compounds in salt marshes. Ecological Processes. 2013;**2**:3

[62] Redzuan NS, Underwood GJC. Movement of microphytobenthos and sediment between mudflats and salt marsh during spring tides. Frontiers in Marine Science. 2020;**7**:496

[63] Victor S, Golbuu Y, Wolanski E, Richmond R. Fine sediment trapping in two mangrove-fringed estuaries exposed to contrasting land-use intensity, Palau, Micronesia. Wetlands Ecology and Management. 2004;**12**:277-283

[64] Victor S, Neth L, Golbuu Y, Wolanski E, Richmond RH. Sedimentation in mangroves and coral reefs in a wet tropical island, Pohnpei, Micronesia. Estuarine Coastal and Shelf Science. 2006;**66**:409-416

[65] Thompson JD. *Spartina anglica*, characteristic feature, or invasive weed of coastal salt marshes? Biologist. 1990;**37**:9-12

[66] Boorman LA, Garbutt A, Barrat D. The role of vegetation in determining patterns of the accretion in salt marsh sediment. In: Sedimentary Processes in the Intertidal Zone. Special Publication No. 139. London: Geological Society; 1998. pp. 389-399

[67] Gray AJ, Marshall DF, Raybould AF. A century of evolution of *Spartina anglica*. Advances in Ecological Research. 1991;**21**:1-62

[68] Langlois E, Bonis A, Bouzille JB. The response of *Puccinellia maritima* to burial: A key to understanding its role in salt marsh dynamics? Journal of Vegetation Science. 2001;**12**:289-297

[69] Langlois E, Bonis A, Bouzille JB. Sediment and plant dynamics in saltmarshes pioneer zone: *Puccinellia maritima* as a key species? Estuarine, Coastal and Shelf Science. 2003;**56**:239-249

[70] Lefeuvre JC, Dame RF. Comparative studies of salt marsh processes on the new and old worlds: An introduction. In: Mitsch WJ, editor. Global Wetlands. Old World and New. Amsterdam: Elsevier; 1994. pp. 169-179

[71] Lefeuvre JC. Effects of Environmental Change on European Salt Marshes. Structure, Functioning and Exchange Potentialities with Marine Coastal Waters. Vol. 5. Rennes: University of Rennes; 1996. p. 50

[72] Cahoon DR, Lynch JC, Powell AN. Marsh vertical accretion in a Southern California estuary. Estuarine, Coastal and Shelf Science. 1996;**43**:19-32

[73] Bishop AW. The use of the slip circle in the stability analysis of slopes. Géotechnique. 1955;**5**:7-17

[74] Gasparotto A, Darby SE, Leyland J, Carling PA. Water level fluctuations drive bank instability in a hypertidal estuary. Earth Surface Dynamics. 2023;**11**:343-361

[75] Feagin RA, Lozada-Bernard SM, Ravens TM, Möller I, Yeager KM, Baird AH. Does vegetation prevent wave erosion of salt marsh edges? Proceedings National Academy of Sciences USA. 2009;**106**:10109-10113

[76] D'Alpaos A, Lanzoni S, Rinaldo A, Marani M. Intertidal

eco-geomorphological dynamics and hydrodynamic circulation. In: Perillo GME, Wolanski E, Cahoon DR, Brinson MM, editors. Coastal Wetlands, an Integrated Ecosystem Approach. Amsterdam: Elsevier; 2009. pp. 159-184

[77] Gedan KB, Kirwan ML, Wolanski E, Barbier E, Silliman BR. The present and future role of coastal wetlands in protecting shorelines: Answering recent challenges to the paradigm. Climatic Change. 2011;**106**:7-29

[78] Chen Y, Thompson CEL, Collins MB. Saltmarsh creek bank stability: Biostabilisation and consolidation with depth. Continental Shelf Research. 2012;**35**:64-74

[79] Chen Y, Thompson C, Collins M. Controls on creek margin stability by the root systems of saltmarsh vegetation, Beaulieu estuary. Southern England. Anthropocene Coasts. 2019;**2**:21-38

[80] Williams D. Recent, rapid evolution of the lower Mary River estuary and flood plains. In: Wolanski E, editor. Estuaries of Australia in 20250 and beyond. Dordrecht: Springer; 2014. pp. 277-287

[81] Ingram HAP, Barclay AM, Coupar AM, Glover JG, Lynch BM, Sprent JI. *Phragmites* performance in reed beds in the Tay estuary. Proceedings of the Royal Society of Edinburgh, Section B: Biological Sciences. 1980;**78**(3-4):89-107

[82] Salomons W, Forstner U. Metals in the Hydrocycle. Berlin: Springer-Verlag; 1984. p. 349

[83] Kirby R. Challenges of restoring polluted industrialised muddy NW European estuaries. In: Wolanski E, Day JW, Elliott M, Ramachandran R, editors. Coasts and Estuaries. The Future. Amsterdam: Elsevier; 2019. pp. 413-425

[84] Cowie G, Woulds C, Cowie G. Tracer studies of benthic communities and biogeochemical processes in coastal and estuarine marine environments. In: Wolanski E, McLusky D, editors. Treatise on Estuarine and Coastal Science. Vol. 4. Amsterdam: Elsevier; 2011. pp. 39-70

[85] Millward GE, Liu YP. Modelling metal desorption kinetics in estuaries. The Science of Total Environment. 2003;**314-316**:613-623

[86] Fitzsimons MF, Lohan MC, Tappin AD, Millward GE. The role of suspended particles in estuarine and coastal biogeochemistry. In: Wolanski E, McLusky D, editors. Treatise on Estuarine and Coastal Science. Vol. 4. Amsterdam: Elsevier; 2011. pp. 71-114

[87] Du Laing G. Redox metal processes and controls in estuaries. In: Wolanski E, McLusky D, editors. Treatise on Estuarine and Coastal Science. Vol. 4. Amsterdam: Elsevier; 2011. pp. 115-141

[88] Varma R, Turner A, Brown MT, Millward GE. Metal accumulation kinetics by the estuarine macroalga, *Fucus ceranoides*. Estuarine, Coastal and Shelf Science. 2013;**128**:33-40

[89] Covelli S. The MIRACLE project: An integrated approach to understanding biogeochemical cycling of mercury and its relationship with lagoon clam farming. Estuarine, Coastal and Shelf Science. 2012;**113**:1-6

[90] Constantin de Magny G, Mozumder PK, Grim CJ, Hasan NA, Naser MN, Alam M, et al. Role of zooplankton diversity in *vibrio cholerae* population dynamics and in the incidence of cholera in the Bangladesh Sundarbans. Applied and Environmental Microbiology. 2011;**77**:6125-6132

[91] Batabyal P, Einsporn MH, Mookerjee S, Palit A, Neogi SB, Nair GB,

et al. Influence of hydrologic and anthropogenic factors on the abundance variability of enteropathogens in the Ganges estuary, a cholera endemic region. Science of the Total Environment. 2014;**472**:154-161

[92] McAnnally WH Jr, Hayter EJ. Estuarine boundary layers and sediment transport. In: Cheng RT, editor. Residual Currents and Long-Term Transport. New York: Springer-Verlag; 1990. pp. 260-275

[93] Mehta AJ, Srinivas R. Observations on the entrainment of fluid mud by shear flow. In: Mehta AJ, editor. Estuarine Cohesive Sediment Transport. Washington D.C.: American Geophysical Union; 1993. pp. 224-246

[94] King B, Wolanski E. Bottom friction reduction in turbid estuaries. In: Mixing in Estuaries and Coastal Seas. Washington D.C.: American Geophysical Union; 1996. pp. 325-337

[95] de Jonge VN, Schuttelaars HM, van Beusekom JEE, Talke SA, de Swart HE. The influence of channel deepening on estuarine turbidity levels and dynamics, as exemplified by the ems estuary. Estuarine, Coastal and Shelf Science. 2014;**139**:46-59

[96] Elliott M, Day JW, Ramachandran R, Wolanski E. A synthesis: What is the future for coasts, estuaries, deltas and other transitional habitats in 2050 and beyond. In: Wolanski E, Day J, Elliott M, Ramachandran R, editors. Coasts and Estuaries. The Future. Amsterdam: Elsevier; 2019. pp. 1-28

[97] Lepage M, Capderrey C, Meire P, Elliott M. Estuarine degradation and rehabilitation. In: Whitfield AK, Able KW, Blaber SJM, Elliott M, editors. Fish and Fisheries in Estuaries – A Global Perspective. Oxford: John Wiley & Sons; 2022. pp. 458-552

[98] Lepage M, Elliott M, Capderrey C, Cabral H. The ecological restoration of estuaries: Protection of people and combating the erosion of biodiversity. In: Salles D, Mainguy G, De Godoy Leski C, editors. Estuarine Cities Facing Global Change: Towards Anticipatory Governance. Oxford: John Wiley & Sons; 2023. pp. 69-23

[99] Spalding M, McIvor A, Beck M, Koch E, Möller I, Reed D, et al. Coastal ecosystems: A critical element of risk reduction. Conservation Letters. 2013;**11**:1-9

[100] Elliott M, Mander L, Mazik K, Simenstad C, Valesini F, Whitfield A, et al. Ecoengineering with ecohydrology: Successes and failures in estuarine restoration. Estuarine, Coastal and Shelf Science. 2016;**176**:12-35

[101] Elliott M, Wither A. Environmental Consequences and Management of Coastal Industries. Amsterdam: Elsevier; 2024. p. 71

[102] McLeod IM, zu Ermgassen PSE, Gillies CL, Hancock B, Humphries A. Can bivalve habitat restoration improve degraded estuaries. In: Wolanski E, Day J, Elliott M, Ramachandran R, editors. Coasts and Estuaries. The Future. Amsterdam: Elsevier; 2019. pp. 417-442

[103] Fernandes LFS, Pinto AAS, Terêncio DPS, Pacheco FAL, Cortes RMV. Combination of ecological engineering procedures applied to morphological stabilization of estuarine banks after dredging. Water. 2020;**12**:391

Chapter 6

Flocculation Dynamics of Cohesive Sediment in Turbulent Flows Using CFD-DEM Approach

Xiao Yu, Sivaramakrishnan Balachandar, Jarrell Smith and Andrew J. Manning

Abstract

Two-phase computational fluid dynamics - discrete element method (CFD-DEM) framework has gained attention in cohesive sediment transport due to its capability of resolving particle-particle interactions and capturing the time evolution of individual flocs and hence the flocculation dynamics of cohesive sediment in turbulent flows. For cohesive sediments of size smaller than the Kolmogorov length scale, the point-particle approach is commonly used, in which the flow around particles is not fully resolved, and the hydrodynamic force on particles is parameterized by the drag law. The accuracy of floc dynamics, aggregation, breakup, and reshaping therefore strongly depends on force parameterization of individual point-particles that make up the floc. In this chapter, we review recent advances in the state-of-art two-phase CFD-DEM model approach on cohesive sediment transport and make recommendation for future research.

Keywords: two-phase approach, CFD-DEM coupling, long-range hydrodynamic interactions, drag model, flocculation dynamics

1. Introduction

Cohesive sediment transport is important for many geoscience and engineering applications. Cohesive sediments can adsorb pollutants (e.g., heavy metals, microplastics, and nutrients) and are a concern for water quality [1, 2]. The transport of estuarine and nearshore cohesive sediment plays an important role in coastal processes and the functioning of healthy ecosystems [3]. Cohesive sediment particles can aggregate with organic particles, such as bacteria, phytoplankton, and algae to form large flocs, marine snow [4], or algal-sediment aggregates [5]. The transport of flocs is therefore one of the key processes in the biogeochemical cycling of the ocean. Modeling the dynamics and transport of suspended sediment is essential to calculate sediment budgets and to provide relevant knowledge for the investigation of biogeochemical cycles. During transport, cohesive sediments undergo a complex flocculation process of aggregation, breakup, and reshaping, which determines the floc size and shape spectrum that in turn affects their sedimentation rate. Our ability to confidently predict the suspended sediment transport critically depends on how accurately we are able to parameterize the shape and size distribution of the cohesive sediments and hence their density and settling velocity.

Population balance equation (PBE) approach has been widely used to model flocculation dynamics in suspended cohesive sediment, in which floc properties change with ambient flow conditions due to either aggregation or breakup processes [6, 7]. By assuming fractal entities of flocs, the aggregation and breakup processes can be modeled as power-law relations. With recent advancements in imaging techniques, Spencer et al. [8] proposed a five-level hierarchy of floc aggregation, in which the floc structures are scale dependent. For microflocs ($O(10\,\mu m)$), electrochemical forces and bridging mechanism by biological activities are of equal importance. For macroflocs ($O(100\ \mu m$–$1\ mm)$), the backbone mechanism due to protruding filamentous bacteria influences the size and shape of macroflocs. The filaments can provide structural integrity by increasing the tensile strength of the flocs. Through the anchoring mechanism, the abundance and morphotype of bacteria affect the shapes of megaflocs ($O(1\ mm)$). Due to the scale dependence of flocs structures, fractal theory cannot represent or predict the behavior of flocs in natural environments. A model framework that can resolve interactions among particles to better characterize the floc structure needs to be developed for more accurate prediction of floc properties in engineering or geoscience applications.

The advantage of the PBE approach is that it can scale to large systems of practical interest. In addition, PBE approach can capture the complex behavior of floc dynamics and predict the bimodal or multimodal floc size distributions that have been observed in the field [9, 10], compared to the single-class model using averaged floc size. However, the accuracy of the PBE approach entirely relies upon the fidelity of the closure models of aggregation and breakup processes, which have generally been motivated by experimental measurements of floc size and settling velocity. The video camera systems used in previous studies also have difficult in tracking individual flocs and directly observe aggregation and breakup processes in 3D [11, 12]. Numerical simulations using CFD-DEM approach can fill this gap to develop accurately models floc-floc and flow-floc interactions and hence improve PBE model accuracy on the prediction of cohesive sediment transport.

The two-phase CFD-DEM coupling framework has gained popularity in studies of sediment transport [13, 14], where sediment particles are modeled as dispersed phase by the discrete element method (DEM) [15, 16]. Turbulence-sediment interactions are modeled by momentum transfer between the continuous fluid phase and the dispersed sediment phase. Two approaches, namely the particle-resolved (PR) and the point-particle (PP) approaches, have been widely used. The PR simulations resolve the flow around individual particles and have been used to investigate important mechanisms in flocculation dynamics, such as aggregation and differential settling [17, 18]. The PR approach is computationally expensive and limited to simulations with a small number of particles (a few thousand), which may not be sufficient to capture mesoscale processes and obtain converged statistics. On the other hand, the PP approach can efficiently implement millions of particles [19]. However, important hydrodynamic interactions at the microscale, such as flow blockage and sheltering by neighboring particles, are not resolved [20].

In this chapter, we will review the recent advancements in the two-phase CFD-DEM approach to cohesive sediment transport studies. The review is structured as follows: Section 2 outlines the fundamental methodology of the CFD-DEM approach, followed by a discussion of its applications to flow modeling and future research (Section 3). We provide a summary of this approach in Section 4.

2. Two-phase CFD-DEM coupling approach for cohesive sediment transport in turbulent flows

2.1 Two-phase modeling approach

Two-phase modeling approach has been widely used in sediment transport. Depending on the volumetric concentration of sediment (ϕ) and Stokes number (St) defined as the ratio of the particle response time to the characteristic timescale of the turbulent flow, different computational approaches has been developed to turbulent multiphase flow [21], including the dusty gas, equilibrium Eulerian, Eulerian, Lagrangian point-particle, and fully resolve approaches. For dilute flow, one-way coupled or two-way coupled approaches are often adopted. In one-way coupled approach, only the effects of the continuous fluid phase on particles are considered, while the reverse effect is neglected. On the other hand, in two-way coupled approach, the fluid-particle interaction term is also introduced in the fluid momentum equation. When interactions among particles (collisions) become important, four-way coupled approach should be used. For sufficiently dilute flow, the fluid does not feel the presence of particles and particles move in response to the fluid; one-way coupling can be used. For large particles of $St > 1$, point-particle Lagrangian approach is uniquely suited. For small particles, the Eulerian approaches are more computationally efficient. In the limit of $St < 0.2$, the equilibrium Eulerian approximation (or mixture theory) can be adopted to study fine sediment suspension [22, 23]. For $0.2 \leq St \leq O(1)$, both the Eulerian two-fluid and Lagrangian point-particle approaches are comparable. For cohesive sediment transport, the volumetric concentration is typically dilute and particle-particle interactions are important; the point-particle Lagrangian approach is better suited.

Two-phase CFD-DEM approach has gained popularity in flocculation dynamics of cohesive sediment in turbulent flows [17, 18, 24–26]. For primary particles of size smaller than the Kolmogorov length scale (η), the local average method of Anderson and Jackson [27] can be used to directly derive the fluid momentum equation as follows (equation set I in [28]):

$$\frac{\partial \alpha}{\partial t} + \nabla \cdot (\alpha \mathbf{u}) = 0, \tag{1}$$

and

$$\frac{\partial \alpha \mathbf{u}}{\partial t} + \nabla \cdot (\alpha \mathbf{u}\mathbf{u}) = -\frac{1}{\rho}\nabla p + \nabla \cdot \boldsymbol{\tau} + \mathbf{F}_{inter} + \mathbf{F}_t, \tag{2}$$

where $\mathbf{u}$ is the fluid velocity, α is the fluid volume fraction, $\boldsymbol{\tau}$ is the viscous stress tensor, and p is the fluid-phase pressure. $\mathbf{F}_{inter}$ is the interphase force exchange term, and $\mathbf{F}_t$ is the external force to drive the turbulent flow. For homogeneous isotropic turbulence, a linear forcing term is commonly implemented [29]. The particle-fluid interaction force is the sum of all particle-fluid interaction forces acting on individual particles, including the drag force, pressure gradient force, viscous force due to fluid shear stress, virtual mass force, Basset force, and lift force [28]. In one-way coupled approach, $\mathbf{F}_{inter}$ term is neglected.

Figure 1a shows the CFD-DEM simulation results of flocculation dynamics of cohesive sediment in homogeneous isotropic turbulence. Initially, 200,000 mono-dispersed spherical particles were introduced into the flow field. Due to collision, flocs formed and grew in size till an equilibrium floc size distribution developed when the aggregation process balanced with the breakup process. The isosurfaces represent the turbulent coherent structures. By turning off the linear forcing term, the decaying turbulent flow relaminarized to quiescent flow. **Figure 1b** shows the gravitational settling of flocs in still water with gravitational acceleration pointing downward. Preferential orientation occurred with major axis of flocs aligning with the direction of gravity for large flocs. In CFD-DEM simulations, sediment is modeled as dispersed phase, and the evolution of floc properties can be directly obtained from model results (**Figure 1b**). Flocs can be identified by the connectivity of spheres, and the floc size can be characterized by the number of primary particles (spheres) consisting the floc. Floc structure can then be derived from the relative position of particles to the center of the floc, defined as

$$\mathbf{x}_f = \frac{\sum_{i=1}^{N} \mathbf{x}_{p,i}}{N}, \tag{3}$$

where $\mathbf{x}_f$ is the center of the floc, $\mathbf{x}_p$ is the center of the particles consisting the floc, i is the label of the particle, and N is the total number of the particles in the floc. To characterize the structure of flocs, fractal dimension and gyration radius have been commonly used. The fractal dimension N_f is defined as

$$N = \left(\frac{L_f}{L_p}\right)^{n_f}, \tag{4}$$

where L_f is the floc size, defined as the floc size along the longest principle axis, and L_p is the size of the primary particle.

The gyration radius is defined as the root-mean-square distance of particles from the floc center as

$$R_g = \sqrt{\frac{1}{N} \sum_{i=1}^{N} \left|\mathbf{x}_{p,i} - \mathbf{x}_f\right|^2}. \tag{5}$$

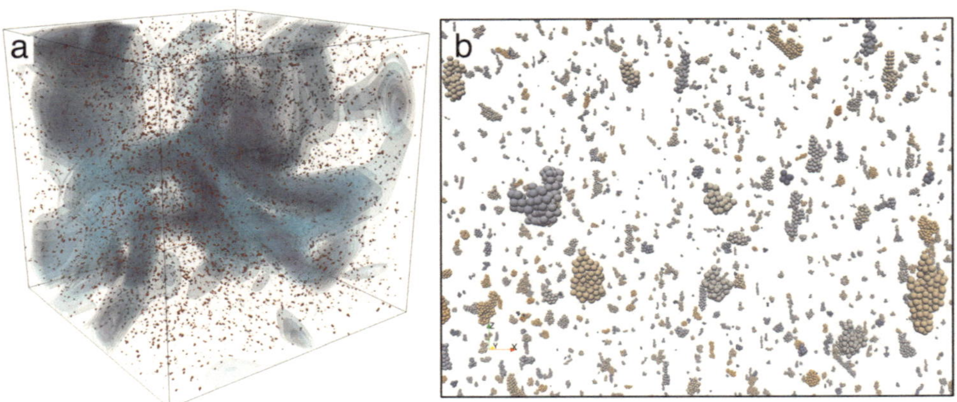

Figure 1.
(a) Flocculation dynamics of cohesive sediment in homogeneous isotropic turbulence. (b) Preferential orientation of flocs with respect to gravity in gravitational sedimentation with effects of long-range hydrodynamic interactions.

By turning off the linear forcing term, turbulence was allowed to decay over time, and the flow relaminarized to a quiescent state. **Figure 1b** shows the gravitational settling of flocs in the final still water with gravitational acceleration pointing downward. Preferential orientation occurred with the major axis of flocs aligning with the direction of gravity for large flocs.

2.2 Discrete element method for cohesive sediment

Interactions among primary particles play an important role in aggregation and breakup processes of cohesive sediment. To resolve particle-particle interactions, the sediment phase is modeled by the discrete element method (DEM), in which motions of individual particles are tracked following

$$\frac{d\mathbf{x}_i}{dt} = \mathbf{v}_i, \tag{6}$$

$$m_i \frac{d\mathbf{v}_i}{dt} = \sum \mathbf{F}_i, \tag{7}$$

and

$$I_i \frac{d\omega_\mathbf{i}}{dt} = \sum \mathbf{T}_i. \tag{8}$$

where $\mathbf{x}$ is the position vector, $\mathbf{v}$ is the particle velocity vector, $\mathbf{F}$ is the force vector, and m is the mass of the particle. The subscript "i" is the particle label. The force on particle "i" is the sum of the collision force ($\mathbf{F}_c$) between particle i and all other particles j, the hydrodynamic force ($\mathbf{F}_f$), and the gravitational force ($\mathbf{F}_g$) as $\mathbf{F}_i = \sum_j \mathbf{F}_{c,ij} + \mathbf{F}_{f,i} + \mathbf{F}_{g,i}$. I is the moment of inertia, ω is the angular velocity of the particle, and $\mathbf{T}$ is the hydrodynamic torque on the particle.

2.2.1 Cohesive force: DLVO and XDLVO theories

In CFD-DEM approach, the stickiness of cohesive sediment is modeled by pairwise interaction forces, which are given as functions of distance between particles. Parteli et al. [30] showed both types of attractive forces, namely, adhesion and nonbonded van der Waals forces, need to be taken into account to accurately model the particle interactions [30]. DLVO theory [31, 32] has been widely used in cohesive sediment studies to model the electrochemical interactions among particles, in which nonbonded attractive van der Waals and repulsive electrostatic double layer interactions are considered. In CFD-DEM simulations, the repulsive electrostatic interactions are often neglected and the van der Waals force is given by

$$\mathbf{F}_{vdW} = \begin{cases} \dfrac{A_H R_{eff}}{6D_{min}^2}\mathbf{e}_n & \text{if} \quad \delta_n > 0, \\ \dfrac{A_H R_{eff}}{6(\delta_n - D_{min})^2}\mathbf{e}_n & \text{if} \quad -D_{max} \le \delta_n \le 0, \\ 0, & \text{if} \quad \delta_n < -D_{max}, \end{cases} \tag{9}$$

in which δ_n is the overlap distance (**Figure 2**), and $\mathbf{e}_n$ is the normal unit vector connecting centers of two spheres. A_H is the Hamaker constant. D_{min} is introduced to avoid the singularity of the equation at $\delta_n = 0$. As the surface of the particle is not smooth, D_{min} can also be interpreted as the surface roughness as there is always a minimal distance between two particles at contact. D_{max} is the cutoff distance of the van der Waals interaction. The effective radius is defined as $R_{eff} = \frac{R_i R_j}{R_i + R_j}$ for two contacting particles of radius R_i and R_j, respectively.

The classic DLVO theory models the electrochemical interactions among particles and assumes the particle surfaces to be chemically inert, which is not valid for organic particles, such as bacteria. As the organic particles play important role in floc properties [8] and hence the flocculation dynamics, effects from microbial communities need to be considered in cohesive sediment transport. To model biological interactions, DLVO theory has been extended as XDLVO [33, 34] to include additional mechanisms, such as bridging effects and acid-base energy from hydrogen bonding [35–38]. DLVO and XDLVO theories provide the ideal framework for modeling the complex cohesive particle interactions. In natural environments, several different mechanisms simultaneously contribute to the attractive force among cohesive sediment particles (i.e., stickiness) and affect the floc properties. To predict flocculation dynamics more accurately, measurement on particle interaction forces becomes important. With Atomic Force Microscopy (AFM), particle level force as a function of distance between two particles can be measured and directly implemented in DEM simulations [39–41]. However, detailed measurements on particle interaction force are still limited; systematic studies to quantify the interaction forces between particles under different ionic strengths, surface coasting, pH, biomaterials, and so forth need to be conducted.

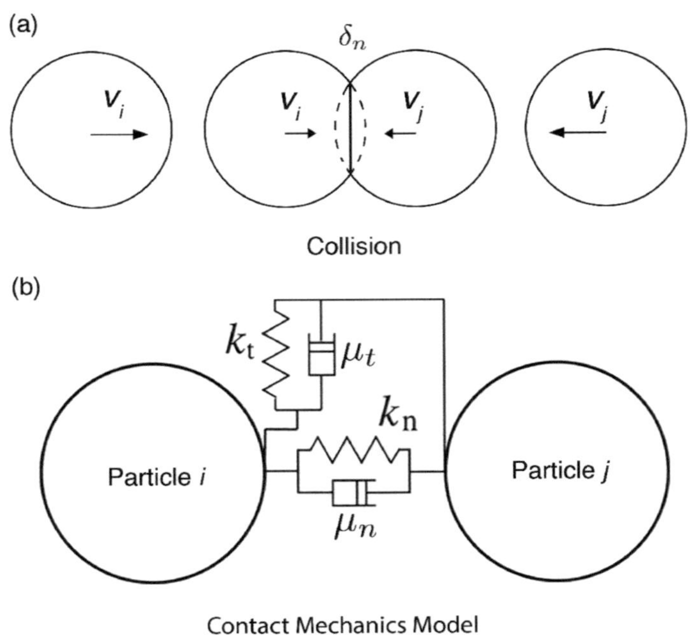

Figure 2.
(a) Schematic of a particle-particle collision handled by the soft-sphere discrete element method. The overlap distance δ_n represents the deformation in the contact region. (b) The spring-dashpot soft-sphere contact model in discrete element method.

2.2.2 Adhesive contact mechanics

In DEM framework, sediment particles are modeled as soft spheres, allowing small overlaps between two particles in contact. Contact mechanics models are often used to account for the collisional behavior of particles, which provide a variety of options for the normal, tangential, rolling, and twisting forces resulting from the contact. These forces are hence implemented to model the complex interactions among the cohesive sediment particles. To model adhesion of particles, two conflicting approximations have been developed, namely the Johnson-Kendall-Roberts (JKR) approximation [42] and the Derjaguin-Muller-Toporov (DMT) approximation [43]. In JKR model, adhesion forces outside the area of contact are neglected and elastic stresses at the edge of the contact are infinite. In DMT model, the geometry is assumed to be Hertzian and hence the adhesion force could not deform the surface even the adhesion is taken into account [44]. The dispute was about the magnitude of the pull-off force. Tabor [45] introduced a parameter μ [45], such that for small μ, DMT should be used and for large μ, JKR should be used. Recently, Greenwood [46] pointed out the errors in DMT thermodynamic method and argued that Hertzian geometry does not occur [46]. Therefore, we will focus on the JKR model in the following discussion.

For soft clay particles, the Johnson-Kendall-Roberts (JKR) model gives

$$\mathbf{F}_{ne,jkr} = \left(\frac{4Ea^3}{3R} - 2\pi a^2 \sqrt{\frac{4\gamma E}{\pi a}} \right) \mathbf{e}_n, \tag{10}$$

where a is the radius of the contact zone and is related to the overlap δ_n according to

$$\delta_n = \frac{a^2}{R} - 2\sqrt{\frac{\pi \gamma a}{E}}, \tag{11}$$

in which E is the Young's Modulus, R is the radius of the particle, and γ is the surface energy density. The overlap between particle "i" and particle "j" is given as $\delta_n = R_i + R_j - |\mathbf{x}_i - \mathbf{x}_j|$. JKR model allows for a tensile force beyond contact ($\delta_n < 0$), up to a maximum of $3\pi\gamma R$. When two particles are not in contact initially, they will not experience this force until they come into contact ($\delta_n > 0$); then, as they move apart, they experience a tensile force up to $3\pi\gamma R$ till they lose contact. Therefore, this force can be used to define the yield strength of the floc.

In addition to the adhesive elastic force, a viscoelastic damping force model is used

$$\mathbf{F}_{n,d} = -\eta_D \mathbf{V}_n, \tag{12}$$

where η_D is the viscoelastic damping coefficient and $\mathbf{V}_n$ is the relative velocity along the direction of the unit vector $\mathbf{e}_n$. The total normal force is the sum of the adhesive elastic JKR and viscoelastic damping terms

$$\mathbf{F}_n = \mathbf{F}_{ne,kjr} + \mathbf{F}_{n,d}. \tag{13}$$

The tangential force ($\mathbf{F}_t$) is given by the Mindlin no-slip model [47] as

$$\mathbf{F}_t = -\min(\mu_t F_{n0}, |-k_t a \boldsymbol{\xi} + \mathbf{F}_{t,d}|)\mathbf{e}_t, \tag{14}$$

in which μ_t is the friction coefficient, k_t is the elastic constant for tangential contact, and $\boldsymbol{\xi}$ is the tangential displacement accumulated during the entire duration of the contact. The unit vector $\mathbf{e}_t$ is in the relative tangential velocity direction. $\mathbf{F}_{t,d}$ is the damping term for the tangential force, which follows the same general form as the normal damping force (Eq. (13)) but uses the relative velocity along the direction of the tangential vector $\mathbf{e}_t$. The normal force value F_{n0} used to compute the critical force is given as

$$F_{n0} = |\mathbf{F}_n + 2\mathbf{F}_{pulloff}| = |\mathbf{F}_{ne,jkr} + 6\pi\gamma R|. \tag{15}$$

Cohesive sediment flocs experience restructuring in turbulent flow, in which the relative position of primary particles consisting the flocs changes without breakup [48]. To account for floc restructuring, a rolling friction model of a pseudo-force formulation [49] is often implemented. The rolling pseudo-force model does not contribute to the total force on either particle but acts only to induce an equal and opposite torque on the particle, which allows adjustment of rolling displacement of the contacting pair and hence floc restructuring. The rolling friction model allows the adjustment of rolling displacement of the contacting pair. The rolling pseudo-force is computed analogously to the tangential force as

$$\mathbf{F}_{roll,0} = k_{roll}\boldsymbol{\xi}_{roll} - \gamma_{roll}\mathbf{v}_{roll}, \tag{16}$$

in which k_{roll} is the elastic constant for rolling, γ_{roll} is the damping constant for rolling, $\boldsymbol{\xi}_{roll}$ is the rolling displacement, and $\mathbf{v}_{roll}$ is the relative rolling velocity [50].

A Coulomb friction criterion truncates the rolling pseudo-force if it exceeds a critical value:

$$\mathbf{F}_{roll} = \min(\mu_{roll}F_{n0}, |\mathbf{F}_{roll,0}|)\mathbf{e}_k, \tag{17}$$

where $\mathbf{e}_k$ is the direction of the pseudo-force. The rolling pseudo-force does not contribute to the total force on either particle, but acts only to induce an equal and opposite torque on each particle according to

$$\mathbf{T}_{roll,i} = \frac{R_i R_j}{R_i + R_j}\mathbf{n} \times \mathbf{F}_{roll}, \tag{18}$$

$$\mathbf{T}_{roll,j} = -\mathbf{T}_{roll,i}. \tag{19}$$

In DEM simulations, the time step is constrained by Young's modulus of particles. To accelerate the simulation, particle stiffness is often reduced by several orders of magnitude. Numerous studies show stiffness can be reduced by orders of magnitude without altering collisional behavior of particles [51] and bulk transport rate of sediment [52]. With adhesive JKR model, the adhesive forces need to be scaled appropriately with a reduced stiffness [53, 54]. In addition, in point-particle Lagrangian approach, the flow around particles is not resolved; the lubrication force can affect the collisional behavior of the particles by decelerating two approaching particles. The enhanced dissipation from the lubrication forces can be modeled by a reduced effective restitution coefficient [55], which is defined as the ratio of the magnitude of relative velocity of particles after collision to the magnitude of the velocity before collision.

2.2.3 Hydrodynamic force: long-range hydrodynamic interactions

Maxey-Riley-Gatignol equation describes the motion of a small rigid sphere in a nonuniform flow [56] and is widely used in point-particle Lagrangian approach. The hydrodynamic force on a particle moving in the flow is the sum of the quasi-steady drag force, undisturbed flow force (sum of pressure gradient force and viscous force due to undisturbed fluid shear stress), added mass force, Basset history force, and lift force. For cohesive sediment transport studies, the Stokes number of the flocs is generally small. In most previous studies, only the drag force and pressure gradient force are considered for simplicity. The total hydrodynamic force on particle "i" is therefore given as

$$\mathbf{F}_{hd,i} = \mathbf{F}_{d,i} + \mathbf{F}_{p,i}, \tag{20}$$

in which $\mathbf{F}_d$ and $\mathbf{F}_p$ are the quasi-steady drag force and undisturbed flow force, respectively.

The undisturbed flow force experienced by the particle is calculated as

$$\mathbf{F}_{p,i} = (-\nabla p + \nabla \cdot \tau) V_{p,i}, \tag{21}$$

where the pressure gradient and stress divergence are interpolated to the particle center. In the current formulation, only the hydrostatic pressure is used to calculate the force F_p for simplicity. The buoyancy force due to the hydrostatic pressure is $\mathbf{F}_{p,i} = -\rho_f \mathbf{g} V_{p,i}$, where $\mathbf{g}$ is the gravitational acceleration vector and $V_p = \pi D_p^3/6$ is the volume of the particle.

The accuracy of model prediction on particle motion strongly depends on the drag model. In the context of cohesive sediments, the Reynolds number of the primary particle (2–6 μm) and flocculi (6–50 μm) is smaller than 1. In a turbulent flow, the Reynolds number of aggregates that are smaller than the Kolmogorov scale is expected to be smaller than unity [57, 58]. Since most flocs tend to be smaller than the Kolmogorov scale, the hydrodynamic drag will be evaluated based on Stokes flow approximation for cohesive sediment transport.

The Free Draining Approximation (FDA) has been widely used to study deformation and breakup of aggregates in shear flow due to its simplicity and computational efficiency [59, 60]. In FDA, Stokes drag law is used to calculate the hydrodynamic force of individual particles that make up the floc, neglecting the hydrodynamic interactions between particles within the floc. Previous studies found FDA can yield qualitatively the same results as studies that fully resolve the flow around particles [61, 62]. In FDA, the shielding effect of neighbors is neglected, and hence, the drag force is overpredicted. However, it can be expected that the drag force on a particle will be influenced by its neighbors—those that are attached to it as part of the floc and those that are part of a nearby floc in case of floc-floc interaction. These effects are currently ignored, and their importance must be carefully examined.

The drag force $\mathbf{F}_d$ on particle "i" is given as

$$\mathbf{F}_{d,i} = \frac{1}{2}\rho C_D A |\mathbf{u} - \mathbf{v}_i|(\mathbf{u} - \mathbf{v}_i). \tag{22}$$

ρ is the fluid density, and $A = \pi D_p^2/4$ is the projected area of the spherical particle with D_p as the diameter of the spherical particle. For very dilute flow with sediment concentration $\phi \ll 0.1\%$, the standard drag coefficient C_D for an individual particle is used, which is given as

$$C_D = \begin{cases} \frac{24}{Re_p}\left(1 + 0.15 Re_p^{0.687}\right), & \text{if} \quad Re_p < 1000 \\ 0.44, & \text{otherwise,} \end{cases} \tag{23}$$

where $Re_p = |\mathbf{u} - \mathbf{v}| D_p / \nu$ is the particle Reynolds number, which is mostly likely to be quite small.

In FDA approach, effects of neighbor particles are oversimplified by a correction to the drag coefficient, which is a function of volumetric concentration of sediment (ϕ). It has been claimed that this simplification works reasonably for sand transport with moderate particle Reynolds number as hindered settling. However, in the Stokes limit, the drag force on two nearby particles is smaller than the Stokes drag of a single particle due to the shielding effect [63]. The shielding effect is represented by $\lambda_s = 1 - \lambda$, with $\lambda = F_i / 6\pi\mu a U_\infty$, and F_i is the drag force on the particle.

While a particle-resolved simulation is computationally out of reach, traditional point-particle approach with free draining assumption completely ignores the shielding effect. Different approaches have been developed to model the shielding effects by neighbor particles. At the level of two approaching particles in an unbounded fluid, their hydrodynamic interaction can be completely described in terms resistance functions [64]. For a system of N interacting particles, Brady [65, 66] introduced Stokesian Dynamics (SD) and accelerated Stokesian Dynamics (ASD), which have been applied to study breakage of colloidal aggregates in shear flow [67, 68]. In SD approach, the particles are modeled as rigid spheres, and long-range hydrodynamic interactions are modeled accurately except for close-contact of particles. Stokesian Dynamics simulations are computationally expansive as the computer time scales with total number of particles as N^3. SD approach has only been applied to study aggregation, breakup, and restructuring of aggregates (flocs) under simple shear in laminar flow.

There have been many theoretical studies on the sheltering effects by neighbor particles at Stokes limit. Filippov developed an exact approach using spherical harmonics to calculate the hydrodynamic force and torque on aggregates of spherical particles in the Stokes limit [69]. However, the exact method of Filippov is also computationally expensive and scales as N^3 for a floc containing N particles. On the other hand, data-driven approach has been quite successful in predicting the shielding effects by neighbor particles. Kim and Lee [70] developed an empirical formula for the shielding effect of neighbor particles, which unlike Filippov's exact approach does not require solving a large linear system. Ma et al. [71] developed binary (pairwise) and trinary interaction models enabled by graph neural network to accurately account for the effects of neighboring particles in a computationally efficient manner.

Figure 3 shows numerical simulation results using two different drag models. By including the shielding effect from neighbor particles, the Kim-Lee (K-L) model is able to predict the dependence of floc settling velocity as a function of floc size (represented as the number of primary particles n_f consisting the floc), while the floc settling velocity is constant (single particle settling velocity) in the FDA approach. In addition, the shielding effect from neighbor particles can also affect the breakup and restructuring by turbulence, as it shields the particle from external flow, allowing the growth of larger flocs. The equilibrium floc size distributions show a significant

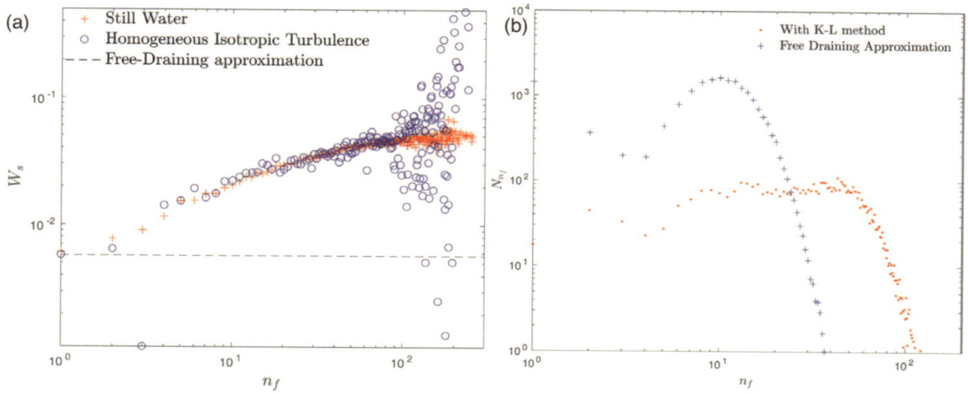

Figure 3.
(a) Floc settling velocity. (b) Floc size distribution.

difference. The K-L model predicts a bimodal distribution with two peaks, while the FDA model predicts a single peak. The first peak appears at around $n_f = 10$ in both cases, while a second peak shows at n_f around 50 with K-L model.

3. Applications and future research

The advantage of the point-particle Lagrangian approach is its capability to resolve interactions among particles, which makes it the ideal candidate to study the flocculation dynamics of cohesive sediment. In many previous studies, the particle-unresolved DNS-DEM approach has been implemented, in which all scales up to the Kolmogorov length scale are fully resolved; however, flows around individual particles are not resolved. Zhao et al. applied the unresolved DNS-DEM model to study flocculation dynamics under homogeneous isotropic turbulence and identified the equilibrium floc size distribution depends on the cohesive number, which is defined as the ratio of floc strength to the turbulence intensity [72]. They also proposed a new flocculation model with a variable fractal dimension that can predict the temporal evolution of the floc size and shape. Zhao et al. [73] applied the one-way coupled CFD-DEM model framework and found that the intermediate shear gives rise to the largest flocs and proposed a simple model for the equilibrium floc size [73]. Yu et al. [24] applied the one-way coupled CFD-DEM model to investigate the reshaping of flocs by turbulence, which leads to more compact flocs at equilibrium [48]. They identified the cohesion number defined as the ratio of floc strength to the turbulent shear rate controls the reshaping mechanisms. With low cohesion number, the breakup-regrowth dominates the reshaping of turbulence dominates as turbulent shear can effectively break the flocs. With high cohesion number, flocs are resilient to small-scale turbulent shear; floc restructuring by hydrodynamic drag plays an important role, in which flocs reshape without breakup. The one-way coupled simulation results have also been used to develop aggregation and breakup kernels [24, 48] by tracking the time evolution of the flocs (**Figure 4**). Yao and Capecelatro [26] performed two-way coupled model simulations to study the breakup of aggregates in homogeneous isotropic turbulence [26]. They also showed the cohesion number plays a key role in the breakup process and conducted a scaling analysis to develop a phenomenological model of the breakup process.

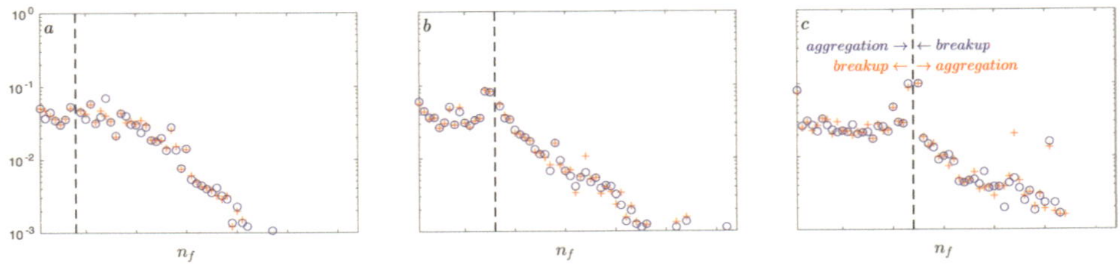

Figure 4.
(a) The probability density function (PDF) that a floc of size n_f becomes a floc of size $n_f = 8$ in a nondimensional time period of 2 (blue circles) and the PDF that a floc of size $n_f = 8$ evolves into a floc of size n_f over the same period (red pluses). (b, c) the same as frame (a) for $n_f = 16$ and 24.

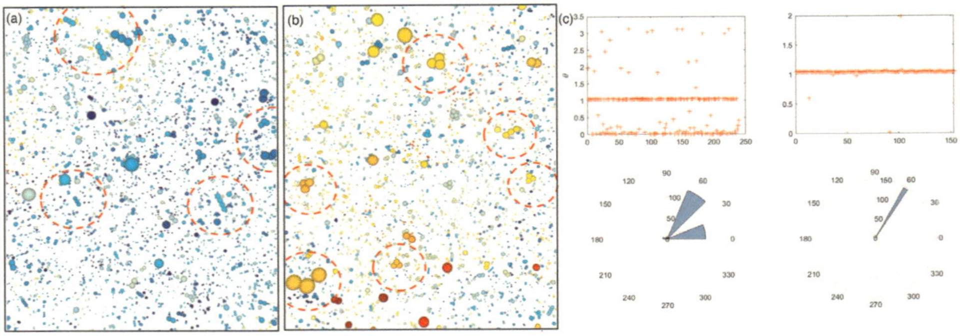

Figure 5.
(a) Flocs at early stage of simulation, (b) flocs at the equilibrium stage, and (c) the corresponding distribution of the interior angles of trimers (a floc consisting of three primary particles).

CFD-DEM approach tracks motions of individual particles; therefore, we can study how turbulence affects floc structures. Different geometric measures have been used to characterize the floc structure, including fractal dimension, gyration radius, and sphericity. Previous studies [24, 48] have identified the compaction of flocs in turbulent flow, in which the gyration radius decreases and fractal dimension increases for flocs of given size (defined by the number of particles in the floc). For instance, the most stable structure for trimers (flocs consisting of three particles) is the equilateral triangle (**Figure 5**) with the interior angle of the triangle formed by three particles of $\pi/3$. At early stage of the simulation, there were many linear trimers with interior angle of 0 or π; however, we observed the dominant peak at $\pi/3$ at the equilibrium stage, suggesting the trimers evolved into the most stable and compact structure by turbulence.

Looking at the future, there is great potential in point-particle Lagrangian modeling framework that can resolve the complex particle-particle and turbulence-particle interactions. More observational data sets on force versus distance between particles are needed to accurately model interactions among particles, including effects of temperature, ionic strength, bacterial strength, and so on. Previous studies did not consider the long-range hydrodynamic interactions among particles mainly due to computational cost. Efficient algorithms to account for the sheltering effects from neighbor particles need to be developed. With both numerical simulation results and observational data sets, detailed scaling analyses need to be conducted to develop parameterization of important flocculation processes, including aggregation, breakup, and restructuring processes. However, we must recognize the role of high-resolution laboratory experiments as a ground truth for model validation, as well as the role of theoretical frameworks to develop more accurate phenomenological models of flocculation dynamics with more accurate representations of physical processes.

4. Conclusion

Two-phase CFD-DEM framework has gained traction in cohesive sediment transport due to its capability of resolving particle-particle interactions. In this chapter, we have presented an overview on recent advances in the state-of-art two-phase CFD-DEM approach on cohesive sediment transport in turbulent environments. We outlined the fundamental methodology of CFD-DEM approach, including models for the CFD-DEM coupling and interparticle interactions, which is followed by a quick discussions of the applications to cohesive sediment transport. The two-phase CFD-DEM approach tracks motions of individual particles and hence can provide valuable information of floc evolution under turbulent flow. These information can be used to develop more accurate parameterizations of aggregation and breakup models for population balance approach for more precise prediction of cohesive sediment transport in natural environments. At last, we identified research gaps on the drag force parameterization, which could guide future research. To overcome the flaws in the widely used free-draining approximation, shielding effects from neighbor particles need to be considered.

Acknowledgements

XY gratefully acknowledges supports by the U.S. Army Engineer Research and Development Center (W912HZ-21-2-0035).

Author details

Xiao Yu[1*†], Sivaramakrishnan Balachandar[1†], Jarrell Smith[2] and Andrew J. Manning[1,3]

1 University of Florida, Gainesville, Florida, US

2 The U.S. Army Engineer Research and Development Center, Vicksburg, Mississippi, US

3 HR Wallingford, Wallingford, UK

*Address all correspondence to: xiao.yu@essie.ufl.edu

† These authors contributed equally.

References

[1] Ongley ED, Krishnappan BG, Droppo G, Rao SS, Maguire RJ. Cohesive sediment transport: Emerging issues for toxic chemical management. Hydrobiologia. 1992;**235**(1):177-187

[2] Chung EG, Fabian AB, Schladow SG. Modeling linkages between sediment resuspension and water quality in a shallow, eutrophic, wind-exposed lake. Ecological Modelling. 2009;**220**(9–10): 1251-1265

[3] Uncles RJ, Joint I, Stephens JA. Transport and retention of suspended particulate matter and bacteria in the Humber-Ouse estuary, United Kingdom, and their relationship to hypoxia and anoxia. Estuaries. 1998;**21**(4):597-612

[4] Alldredge AL, Silver MW. Characteristics, dynamics and significance of marine snow. Progress in Oceanography. 1988;**20**:41-82

[5] Deng Z, He Q, Safar Z, Chassagne C. The role of algae in fine sediment flocculation: In-situ and laboratory measurements. Marine Geology. 2019; **413**:71-84

[6] Lee BJ, Toorman E, Molz FJ, Wang J. A two-class population balance equation yielding bimodal flocculation of marine or estuarine sediments. Water Research. 2011;**45**(5):2131-2145

[7] Son M, Hsu TJ. The effects of flocculation and bed erodibility on modeling cohesive sediment resuspension. Journal of Geophysical Research: Oceans. 2011;**116**(C3):1-18

[8] Spencer KL, Wheatland JA, Bushby AJ, Carr SJ, Droppo IG, Manning AJ. A structure–function based approach to floc hierarchy and evidence for the non-fractal nature of natural sediment flocs. Scientific Reports. 2021; **11**(1):1-10

[9] Eisma D. Flocculation and de-flocculation of suspended matter in estuaries. Netherlands Journal of Sea Research. 1986;**20**(2–3):183-199

[10] Lee BJ, Fettweis M, Toorman E, Molz FJ. Multimodality of a particle size distribution of cohesive suspended particulate matters in a coastal zone. Journal of Geophysical Research: Oceans. 2012;**117**(C3):1-17

[11] Blaser S. Flocs in shear and strain flows. Journal of Colloid and Interface Science. 2000;**225**(2):273-284

[12] Manning AJ, Dyer KR, Lafite R, Mikes D. Flocculation measured by video based instruments in the Gironde estuary during the European Commission SWAMIEE project. Journal of Coastal Research. 2004;**SI 41**:58-69

[13] Drake TG, Calantoni J. Discrete particle model for sheet flow sediment transport in the nearshore. Journal of Geophysical Research: Oceans. 2001;**106** (C9):19859-19868

[14] Sun R, Xiao H, Sun H. Realistic representation of grain shapes in CFD–DEM simulations of sediment transport with a bonded-sphere approach. Advances in Water Resources. 2017;**107**: 421-438

[15] Kozicki J, Donze FV. YADE-OPEN DEM: An open-source software using a discrete element method to simulate granular material. Engineering Computations. 2009;**26**(7):786-805

[16] Li S, Marshall JS, Liu G, Yao Q. Adhesive particulate flow: The discrete-element method and its application in

energy and environmental engineering. Progress in Energy and Combustion Science. 2011;**37**(6):633-668

[17] Zhang JF, Zhang QH. Lattice Boltzmann simulation of the flocculation process of cohesive sediment due to differential settling. Continental Shelf Research. 2011;**31**(10):S94-S105

[18] Sun R, Xiao H, Sun H. Investigating the settling dynamics of cohesive silt particles with particle-resolving simulations. Advances in Water Resources. 2018;**111**:406-422

[19] Zwick D, Balachandar S. A scalable Euler–Lagrange approach for multiphase flow simulation on spectral elements. The International Journal of High Performance Computing Applications. 2020;**34**(3):316-339

[20] Vowinckel B, Withers J, Luzzatto-Fegiz P, Meiburg E. Settling of cohesive sediment: Particle-resolved simulations. Journal of Fluid Mechanics. 2019;**858**:5-44

[21] Balachandar S, Eaton JK. Turbulent dispersed multiphase flow. Annual Review of Fluid Mechanics. 2010;**42**: 111-133

[22] Ozdemir CE, Hsu TJ, Balachandar S. A numerical investigation of fine particle laden flow in an oscillatory channel: The role of particle-induced density stratification. Journal of Fluid Mechanics. 2010;**665**:1-45

[23] Yu X, Ozdemir CE, Hsu TJ, Balachandar S. Numerical investigation of turbulence modulation by sediment-induced stratification and enhanced viscosity in oscillatory flows. Journal of Waterway, Port, Coastal, and Ocean Engineering. 2014;**140**(2):160-172

[24] Yu M, Yu X, Balachandar S. Particle nonresolved DNS-DEM study of flocculation dynamics of cohesive sediment in homogeneous isotropic turbulence. Water Resources Research. 2022;**58**(6):e2021WR030402

[25] Zhao K, Pomes F, Vowinckel B, Hsu TJ, Bai B, Meiburg E. Flocculation of suspended cohesive particles in homogeneous isotropic turbulence. Journal of Fluid Mechanics. 2021;**921**:A17

[26] Yao Y, Capecelatro J. Deagglomeration of cohesive particles by turbulence. Journal of Fluid Mechanics. 2021;**911**:A10

[27] Anderson TB, Jackson R. Fluid mechanical description of fluidized beds. Equations of motion. Industrial & Engineering Chemistry Fundamentals. 1967;**6**(4):527-539

[28] Zhou ZY, Kuang SB, Chu KW, Yu A. Discrete particle simulation of particle–fluid flow: Model formulations and their applicability. Journal of Fluid Mechanics. 2010;**661**:482-510

[29] Rosales C, Meneveau C. Linear forcing in numerical simulations of isotropic turbulence: Physical space implementations and convergence properties. Physics of Fluids. 2005; **17**(9):1-8

[30] Parteli EJ, Schmidt J, Blümel C, Wirth KE, Peukert W, Pöschel T. Attractive particle interaction forces and packing density of fine glass powders. Scientific Reports. 2014;**4**(1):6227

[31] Derjaguin B, Landau L. The theory of stability of highly charged lyophobic sols and coalescence of highly charged particles in electrolyte solutions. Acta Physicochimica URSS. 1941;**14** (633–52):58

[32] Verwey EJW, Overbeek JTG. Theory of the stability of lyophobic colloids.

Journal of Colloid Science. 1955;**10**(2): 224-225

[33] Hermansson M. The DLVO theory in microbial adhesion. Colloids and Surfaces B: Biointerfaces. 1999;**14**(1–4): 105-119

[34] Grasso D, Subramaniam K, Butkus M, Strevett K, Bergendahl J. A review of non-DLVO interactions in environmental colloidal systems. Reviews in Environmental Science and Biotechnology. 2002;**1**:17-38

[35] Van Oss CJ, Good RJ, Chaudhury MK. The role of van der Waals forces and hydrogen bonds in "hydrophobic interactions" between biopolymers and low energy surfaces. Journal of Colloid and Interface Science. 1986;**111**(2):378-390

[36] De Gennes PG. Polymers at an interface; a simplified view. Advances in Colloid and Interface Science. 1987;**27** (3–4):189-209

[37] Ong YL, Razatos A, Georgiou G, Sharma MM. Adhesion forces between E. Coli bacteria and biomaterial surfaces. Langmuir. 1999;**15**(8):2719-2725

[38] Thwala JM, Li M, Wong MC, Kang S, Hoek EM, Mamba BB. Bacteria–polymeric membrane interactions: Atomic force microscopy and XDLVO predictions. Langmuir. 2013;**29**(45): 13773-13782

[39] Zhang J, Zhang Q, Maa JPY, Shen X, Liang J, Yu L, et al. Effects of organic matter on interaction forces between polystyrene microplastics: An experimental study. Science of the Total Environment. 2022;**844**:157186

[40] Tolhurst TJ, Gust G, Paterson DM. The influence of an extracellular polymeric substance (EPS) on cohesive sediment stability. In: Proceedings in Marine Science. Vol. 5. Elsevier; 2002. pp. 409-425

[41] de Kerchove AJ, Elimelech M. Structural growth and viscoelastic properties of adsorbed alginate layers in monovalent and divalent salts. Macromolecules. 2006;**39**(19):6558-6564

[42] Kendall K. The adhesion and surface energy of elastic solids. Journal of Physics D: Applied Physics. 1971;**4**(8):1186

[43] Derjaguin BV, Muller VM, Toporov YP. Effect of contact deformations on the adhesion of particles. Journal of Colloid and Interface Science. 1975;**53**(2):314-326

[44] Maugis D. Adhesion of spheres: The JKR-DMT transition using a Dugdale model. Journal of Colloid and Interface Science. 1992;**150**(1):243-269

[45] Tabor D. Surface forces and surface interactions. In: Plenary and Invited Lectures. U.S.: Academic Press; 1977. pp. 3-14

[46] Greenwood JA. Derjaguin and the DMT theory: A farewell to DMT? Tribology Letters. 2022;**70**(2):61

[47] Mindlin RD. Compliance of Elastic Bodies in Contact1949. pp. 259-268

[48] Yu M, Yu X, Mehta AJ, Manning AJ, Khan F, Balachandar S. Persistent reshaping of cohesive sediment towards stable flocs by turbulence. Scientific Reports. 2023;**13**(1):1760

[49] Luding S. Cohesive, frictional powders: Contact models for tension. Granular Matter. 2008;**10**(4):235-246

[50] Wang Y, Alonso-Marroquin F, Guo WW. Rolling and sliding in 3-D

discrete element models. Particuology. 2015;**23**:49-55

[51] Tsuji Y, Kawaguchi T, Tanaka T. Discrete particle simulation of two-dimensional fluidized bed. Powder Technology. 1993;**77**(1):79-87

[52] Elghannay HA, Tafti DK. Sensitivity of numerical parameters on DEM predictions of sediment transport. Particulate Science and Technology. 2018;**36**(4):438-446

[53] Hærvig J, Kleinhans U, Wieland C, Spliethoff H, Jensen AL, Sørensen K, et al. On the adhesive JKR contact and rolling models for reduced particle stiffness discrete element simulations. Powder Technology. 2017;**319**:472-482

[54] Chen S, Liu W, Li S. A fast adhesive discrete element method for random packings of fine particles. Chemical Engineering Science. 2019;**193**:336-345

[55] Liu G, Yu F, Wang S, Liao P, Zhang W, Han B, et al. Investigation of interstitial fluid effect on the hydrodynamics of granular in liquid-solid fluidized beds with CFD-DEM. Powder Technology. 2017;**322**:353-368

[56] Maxey MR, Riley JJ. Equation of motion for a small rigid sphere in a nonuniform flow. The Physics of Fluids. 1983;**26**(4):883-889

[57] Balachandar S. A scaling analysis for point–particle approaches to turbulent multiphase flows. International Journal of Multiphase Flow. 2009;**35**(9): 801-810

[58] Ling Y, Parmar M, Balachandar S. A scaling analysis of added-mass and history forces and their coupling in dispersed multiphase flows. International Journal of Multiphase Flow. 2013;**57**:102-114

[59] Potanin AA. On the computer simulation of the deformation and breakup of colloidal aggregates in shear flow. Journal of Colloid and Interface Science. 1993;**157**(2):399-410

[60] Higashitani K, Iimura K, Sanda H. Simulation of deformation and breakup of large aggregates in flows of viscous fluids. Chemical Engineering Science. 2001;**56**(9):2927-2938

[61] Becker V, Schlauch E, Behr M, Briesen H. Restructuring of colloidal aggregates in shear flows and limitations of the free-draining approximation. Journal of Colloid and Interface Science. 2009;**339**(2):362-372

[62] Saxena A, Kroll-Rabotin JS, Sanders RS. Numerical investigation of the respective roles of cohesive and hydrodynamic forces in aggregate restructuring under shear flow. Journal of Colloid and Interface Science. 2022; **608**:355-365

[63] Debye P, Bueche AM. Intrinsic viscosity, diffusion, and sedimentation rate of polymers in solution. The Journal of Chemical Physics. 1948;**16**(6): 573-579

[64] Jeffrey DJ, Onishi Y. Calculation of the resistance and mobility functions for two unequal rigid spheres in low-Reynolds-number flow. Journal of Fluid Mechanics. 1984;**139**:261-290

[65] Brady JF, Bossis G. Stokesian dynamics. Annual Review of Fluid Mechanics. 1988;**20**(1):111-157

[66] Sierou A, Brady JF. Accelerated stokesian dynamics simulations. Journal of Fluid Mechanics. 2001;**448**:115-146

[67] Harshe YM, Lattuada M, Soos M. Experimental and modeling study of breakage and restructuring of open and

dense colloidal aggregates. Langmuir. 2011;**27**(10):5739-5752

[68] Harshe YM, Lattuada M. Breakage rate of colloidal aggregates in shear flow through Stokesian dynamics. Langmuir. 2012;**28**(1):283-292

[69] Filippov AV, Zurita M, Rosner DE. Fractal-like aggregates: Relation between morphology and physical properties. Journal of Colloid and Interface Science. 2000;**229**(1):261-273

[70] Kim J, Lee S. Modeling drag force acting on the individual particles in low Reynolds number flow. Powder Technology. 2014;**261**:22-32

[71] Ma Z, Ye Z, Pan W. Fast simulation of particulate suspensions enabled by graph neural network. Computer Methods in Applied Mechanics and Engineering. 2022;**400**:115496

[72] Zhao K, Vowinckel B, Hsu TJ, Köllner T, Bai B, Meiburg E. An efficient cellular flow model for cohesive particle flocculation in turbulence. Journal of Fluid Mechanics. 2020;**889**:R3

[73] Zhao K, Vowinckel B, Hsu TJ, Bai B, Meiburg E. Cohesive sediment: Intermediate shear produces maximum aggregate size. Journal of Fluid Mechanics. 2023;**965**:A5

Chapter 7

Suspended Particle Dynamics during Sand Nourishments and Storm Events: A Comparative Analysis

Daniela Glueck and Hendrik Schubert

Abstract

Understanding sediment transport during sand nourishment and storm events is crucial for coastal management and environmental preservation. This chapter provides key insights for policymakers, coastal engineers, and researchers working to balance coastal protection and ecosystem sustainability. It offers an analysis of suspended particle behavior in the water column post-sand nourishment, a coastal protection measure, in comparison to a natural storm event. This allows estimating spatial and temporal effects of coastal protection measures on turbidity and sedimentation in neighboring areas. Pronounced differences in increased turbidity longevity were observed, with sand nourishment-induced turbidity persisting for at least 3 months, spanning a minimum of 4 km along the coast. Additionally, this study reveals elevated nutrient concentrations in the pore water of nourished sediment and the water column during the sand nourishment, along with an increase in fine sediment particles in the nourished sediment due to the use of sand from sedimentation sinks.

Keywords: suspended particles, Baltic Sea, coastal protection, nutrient release, turbidity

1. Introduction

Coastal protection measures are crucial for safeguarding vulnerable coastal areas, but they pose challenges to environments like sandy beaches and dunes, already threatened by issues such as coastal squeeze [1]. The introduction of foreshore structures, especially hard ones, can impact sediment dynamics, hydrodynamics, and depositional processes [2–4] and influencing the distribution of non-native species [1, 5]. Moreover, protecting cliffs with engineered solutions deactivates them, disrupting sediment supply to beaches and increasing the risk of dune breaches. In areas where dunes serve as the primary coastal protection, preventing breaches is crucial to mitigate the threat of storm surges and floods in the coastal hinterland.

IntechOpen

Hence, enhancement of natural coastal protection is increasingly explored, particularly coastal protection through dunes. Strategies involve mitigating sediment transport, stabilizing the coast addressing negative sediment balances. Sand nourishments, regarded as such nature-based coastal protection measures [6, 7], counter sediment deficits caused by the absence of cliff breaks or sediment traps, ensuring coast-parallel sediment transport preventing shoreline retreat. Well-managed dunes play a crucial role in this protection, although wear and erosion are anticipated and the problem of the sediment deficit, which is erosion, is not solved but only the consequences are covered, necessitating regular nourishments for maintenance [8].

There are three primary methods for nourishing sandy beaches, but all involve using dredged marine sands. The first method uses the sand to replenish the entire intertidal zone's sand budget, while the second and third method aim for installing a sediment budget as reinforcement on the beach and dune, backshore, or in the foreshore area [9, 10]. After deposition of the sand, it is expected that the sediment is redistributed naturally by wave actions and wind [10, 11]. During foreshore nourishments, the sand is mostly dumped or sprayed directly from the ship into the target area while for intertidal or backshore nourishments, the sand needs to be transported onto the beach, for example, with pipelines [10]. However, these marine sediments, characteristic of a marine sea bottom, differ from the nutrient-poor beach and dune ecosystem. Therefore, beach nourishments pose risks by introducing non-native sediments, disturbing natural habitats and dune succession. This includes biota burial and habitat loss [2, 3], along with changes in plant diversity and coverage attributed to significant anthropogenic activity [12, 13], resulting in reduced essential ecosystem functions [14].

But the more fundamental problem is the availability of suitable sand deposits whose sediment properties match the desired backwash sites. The growing demand for larger sand volumes in nourishments poses a potential future challenge: sediment availability. In Mecklenburg-Western Pomerania, the sand volume used for nourishment has steadily increased over the decades (1964–2016), reaching a peak in 2004 with over 2,000,000 thousand cubic meters of sand [15]. As coastal erosion intensifies with rising sea levels, surpassing this volume becomes a plausible scenario (**Figure 1**).

On the other hand, this could be overcome by generating closed cycles, meaning that the material used for sand nourishment is gathered from the areas where eroded material got deposited.

But even doing so, another problem still remains – the disturbance of both, the sand nourishment, as well as the sand extraction ecosystems. Sand nourishment is a massive impact, covering the respective habitat with meters of sand, burying fauna and flora. Especially plants are sensitive because in contrast to fauna, plant diaspores

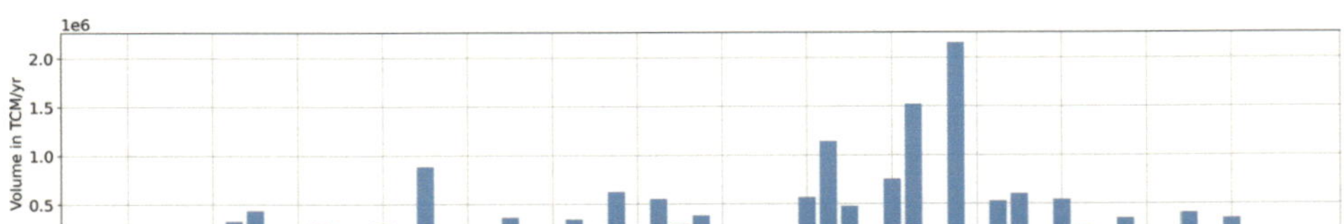

Figure 1.
Volume of sand in thousand cubic meters (TCM) used for sand nourishment in Mecklenburg-Western Pomerania between 1964 and 2016 [15].

and vegetative parts are restricted in their ability to reach back to the surface by elongation in darkness. However, beach vegetation in most of the cases is rather scarce and ephemeral because of trampling by beach visitors [12] and mechanical disturbance during autumn/winter storms with high water levels. Dune vegetation is in a process of succession, which gets interrupted by the nourishment [16]. This loss of plant cover after sand nourishment provokes increased eolian transport, as depicted in **Figure 2**, and resets the succession to the white dune stage [17]. Planting suitable species soon after nourishment is consequently part of such coastal protection measures, but still, the system remains in a white dune stage. But given repeated nourishment events, the succession to brown dune with full coverage or even coastal forest with high resistance to flooding and eolian transport erosion is impossible. Sand nourishment therefore must be seen as a massive impact which restores a desired coastal profile on the cost of biodiversity and erosion resistance provided by vegetation cover.

Figure 2.
Examples of erosion effects and protective measures. Upper left: Undisturbed beach stretch at the Danish coast, fully covered by vegetation; upper right: (planted) dune forest, allowing for complete vegetation cover beneath the trees, at north Jylland; middle left: Eolian sand transport at an unvegetated dune; middle right: Surface of a beach treated by sand nourishment after a storm event (Warnemuende) composed of gravel and coarse sand on fine sand; lower left: Gravel dune at the Atlantic coast; fine material has been blown and washed away irrespective of the availability of fine material at the neighbored beach; lower right: Plumes of fine material eroding from the Dornbusch cliff at Hiddensee.

Moreover, extracting sediments from the donor site during nourishment results in habitat loss. Numerous studies have shown that it takes months to years for conditions at the donor site to return to their pre-extraction state [18–21]. The regeneration timeline is influenced by the extraction method; large-scale dredging creates less deep sediment funnels that replenish more quickly than dredging a limited number of very deep funnels [22, 23].

It must be taken into account that such a reset in dune vegetation succession after a sand nourishment also occurs after storm surges, removing at least part of the dunes [24]. The obvious consequences of sand nourishments for coastal ecosystems, especially increased turbidity and sedimentation in the water column, but also nutrient translocation from the terrestrial part of the beach ecosystem into the aquatic one will occur during such an event too. The same applies for changes in grain size distribution [24] and biodiversity.

So, the question to what extent sand nourishment is comparable to a storm event with respect to duration and strength of mechanical disturbance should be analyzed carefully in order to estimate its overall effect on coastal ecosystems. This applies especially for the Baltic Sea, a microtidal system with pronounced seasonality of storm events [25], where data about the effects of storm surges on nutrient concentration and sedimentation characteristics are scarce.

Our study aims to sort out several key research questions focusing on the dynamics after sediment deposition in coastal areas following sand nourishments in comparison with sediment erosion by storm events:

- By investigating sediment composition at the nourishment site, we seek to understand the behavior of the nourished sediment and sediment sorting processes both in the shallow water as well as on the terrestrial part of the coast after sand nourishment.

- As the nourished sediment was sourced from deeper deposition sites, it may maintain high nutrient concentrations. Nutrient concentrations in both sediment pores and the water column were measured as we intend to estimate the quantity of released nutrients and the temporal and spatial extent of their impact.

- Additionally, we aim to assess the turbidity plume generated by the nourishment, both composition by organic and inorganic particles and its persistence and spatial spread.

- For a meaningful comparison of these findings with natural disturbances, specifically storms, we measured identical parameters there.

This comprehensive approach allows us to gain insights into the short-term and long-term effects of sand nourishments on sediment composition, nutrient release, and turbidity, in order to get a basis for evaluation of the ecological impact of this kind of coastal protection by stakeholders as, for example, environmental agencies, policy makers, and governmental bodies responsible for coastal protection.

The following chapter also follows this order of content after an introduction to the study sites and the selected methods. The results are first discussed at regional level before being placed in a wider context. We conclude with a summary of the results and point out further topics that must be taken into consideration when analyzing sediments in the coastal area of the Baltic Sea.

2. Study sites and methods

The German Baltic Sea Coast is characterized by extensive flat coasts, presenting sandy beaches and shallow water bodies [26]. The originally highly indented coastline was smoothed by shore parallel sediment transport with capacities up to 200,000 m^3 each year [27]. Accumulation of predominantly sandy sedimentary material in calm zones lead to the formation of shallow landforms. Elevated moraine hilltops, remnants of the last ice age, are now connected by narrow land bridges. These land bridges were formed through the erosion of material from the island cores by currents and are now known as the "bodden compensation coast." Moreover, a formation of shallow and sandy landforms with increasing sealing off of former sea bays (= inland Bodden coast) with partly strongly reduced coastal dynamics and sedimentation processes developed [27].

The formation and reshaping of this coastal region is therefore dependent on the sediment supply by erosion of cliffs transported by strong lateral currents. Both sites subject to latest scientific research are located in this area and show this kind of sediment transport but differ regarding their sediment availability (**Figure 3**). Graal Mueritz receives sediment from sand transported from the more western coast sections, while Ahrenshoop originally also depended on more western cliff breaks for sediment supply. But breakwaters now protect some of these cliffs in the 15 km coastal stretch between the cities. A breakwater just before Ahrenshoop serves as an additional sediment trap to stabilize the shore but hinders lateral sediment transport, impacting the adjacent beach. Due to the absence of sediment supply from cliff breaks, the still high lateral sand transport rates weaken the protective function of the dunes in Ahrenshoop, exposing the city to storm surges. The coastal retreat is up to 50 cm per 100 years decreasing the sand budget of the dune and beach. To counteract this, beach nourishments are carried out every few years [8].

Systematic coastal and flood protection at the German Baltic Sea Coast started in the first half of the ninteenth century including dunes stabilized by wooden piles. After more severe storm surges in the second half of the ninteenth and the first half of the twentieth century, these measures were enforced by dikes, followed by groynes, sand nourishments, revetments, and more [27]. The dike in Ahrenshoop was built in

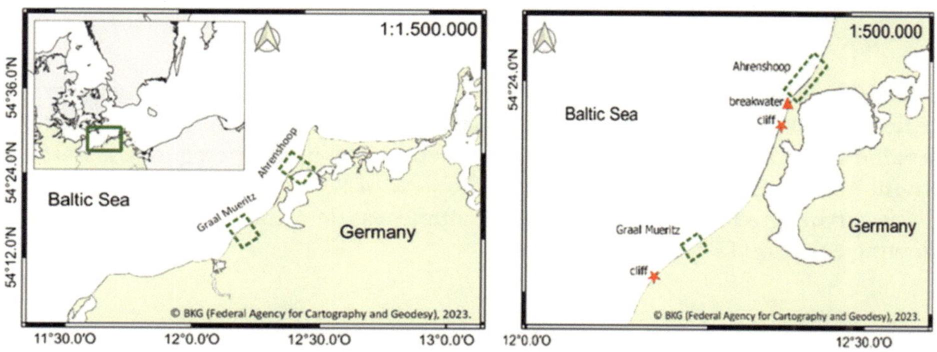

Figure 3.
Map of the nourishment site (Ahrenshoop) and storm site (Graal Mueritz) located at the German Baltic Sea coast in Mecklenburg Western Pomerania in the north-east of Germany. The overview of the Baltic Sea coast with adjacent countries is on the left side. On the right side is the map of the two sites with important landmarks, such as nearby cliffs and breakwater.

1959, the breakwater in 1986, and the groynes in 2006 all support the national coastal protection dune which was already nourished six times in this century [28].

During the most recent nourishment in winter 2021/22, approximately 600 thousand cubic meters of sand were deposited along a 4 km stretch of beach and dune. A duker connected the dredging vessel to the beach, allowing sediment to be flushed through to the beach. Heavy machinery on the beach then distributed the sediment, shaping the new contours and slope of the dune and beach. The desired dune crest height was 5 m with a seaward dune foot height of additional 2 m. The inclination of the seaward slope was 1.3, the beach slope 1.25, and the underwater slope 1.3. A surface protection by planting of more than 50.000 m^2 of marram grass *Ammophila arenaria* ended the measure [28].

In Graal Mueritz, the changes in dune and beach profile are not changing as much as in Ahrenshoop. The natural coastal protection by the coastal dunes in combination with groynes is enough to ensure the safety of the population protected by coastal measures. There are no additional sand nourishments necessary to support the sediment transport toward the beach and dune.

As already mentioned, the distance between the two sites representing exemplary sediment transport conditions is very small with only 15 km. Both sites are located in the southeastern part of the Baltic Sea which is described as a coastal sand accumulation zone consisting of sand bottom based on sand and small gravel [26]. Both sites are subjected to the same wind and wave regime [29], and the hydrological and oceanographical (salinity, temperature) conditions are identical [26]. The sediment characteristics before the nourishment were similar for both mean grain size, 0.3291 ± 0.0883 mm in Ahrenshoop and 0.2778 ± 0.0582 mm in Graal Mueritz, and sorting grade, 1.2589 ± 0.3662 in Ahrenhoop and 1.3011 ± 0.3156 in Graal Mueritz.

For the comparison of the two events at the sampling site, sediment cores and water samples were taken. The upper 20 cm sediment was analyzed for grain size distribution [30] and water content [31], sorting grade [30], and organic and carbonate content [32].

The mean grain size (MGS) of each sediment sample was determined according to the following equation (Eq. (1)) [30]:

$$\mathrm{MGS} = \frac{A * b}{B} * c \tag{1}$$

with A = difference between the two sieve sizes between which the average grain size lies, b = 50 minus the sum of percentages of all fractions below 50%, B = percentage value of the fraction that exceeds 50% when adding up the percentages, starting with the smallest fraction, and c = size of the sieve of B.

The sorting grade(s) of each sediment sample was determined according to the following equation (Eq. (2)) [30]:

$$s = \frac{3 * H + N - 100}{100} \tag{2}$$

with H = percentage of main sediment fraction and N = sum of the main fraction's neighboring fractions.

Suspended particles in the water samples were filtered through a glass fiber filter (pore size = 0.4 µm). The filters were burned at 550°C for 4 h for ignition loss to determine the organic portion. Nutrient concentration was analyzed using the spectrophotometric method by an UV/VIS Spectrometer (Lambda 2, Perkin Elmer) [33]. A pre-experiment showed that surface water samples are sufficient for comparison [29]. The water depth was 4 m at both sites. The water samples in Ahrenshoop were collected from a drifting research vessel. The water samples in Graal Mueritz were taken from a sea bridge that extended about 100 m into the Baltic Sea.

3. Results and discussion

During the nourishment campaign, 600.000 m^3 of sand were deposited along a 4 km long beach stretch at Ahrenshoop. Related to the area (0.4 km^2), this means a cover of about 1.5 m thickness of allochthonous material. As can be seen from **Figure 4**, the nourished beach differs largely from the previous state with respect to grain size distribution. In general, the excavated material had higher proportions of fine material with large differences between the samples taken a feature which could still be observed after 1 year. A distinct difference between beach and dune was the relatively high proportion of fractions >500 µm at the beach compared to the dune before nourishment. This can be explained by preferential erosion of fine material during flooding events at the beach. Consequently, a slight increase in the percentage of coarse material within a year after nourishment can be observed at the beach, but not at the dune.

This assumption is supported by the results of calculation of sorting grade, shown in **Figure 5**. The highest sorting grade was observed at the dune prior to nourishment. Dune systems are built by eolian transport, accumulating mainly highly mobile fine material [34]. Coarse material, as introduced by nourishment, will not be blown away and remains at the surface until being covered by fine material again, a process which seems to take more than the year covered by our observation. Consequently, sorting grade dropped after nourishment and remained low. In this case, one can speculate

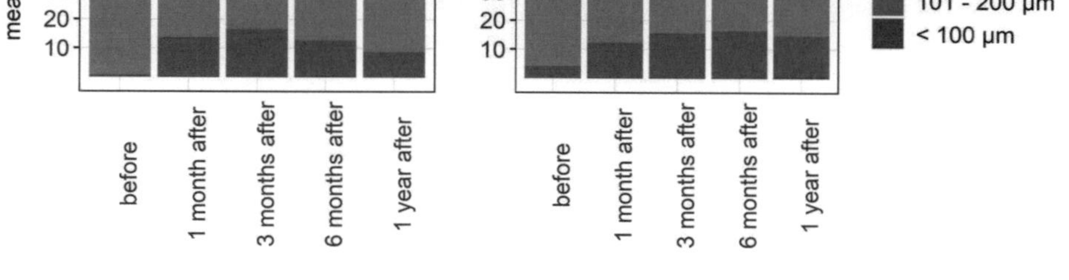

Figure 4.
Grain size composition at the nourishment site. Shown is the percentage of the different grain size classes at the beach (left panel) and dune (right panel) before and after the sand nourishment.

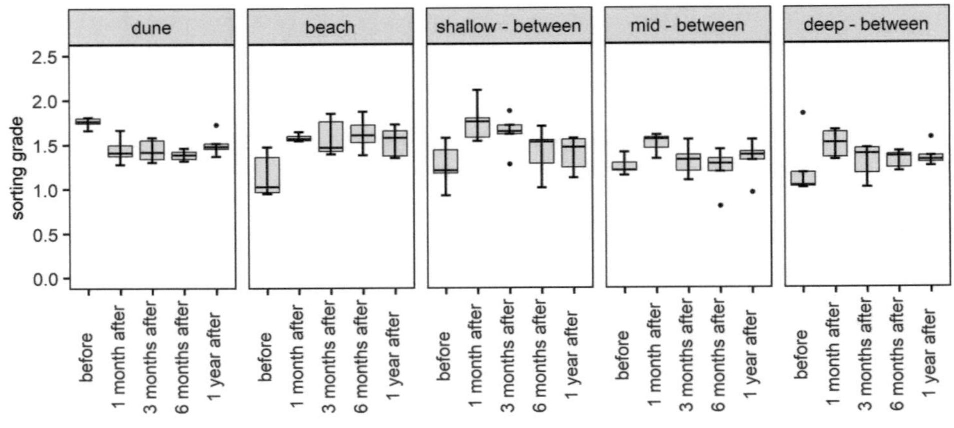

Figure 5.
Sorting grade kinetics at different sections after nourishment. Shown are temporal changes in sorting grade before and after sand nourishment at Ahrenshoop at the dune, the beach, and the midsections between groyne lines in front of the beach ("shallow"), halfway the groyne line ("mid"), and at the end of the groyne line ("deep").

that such a nourishment with coarser fractions may contribute to the resistance of dune systems against eolian erosion [35]. Such reduction of transversal transport may negatively impact leeward areas in a dynamic system similar to the effects described after protection measures against cliff erosion [36].

In contrast to this, sorting grade at the beach and shallow water between the groyne lines increased sharply by the nourishment. All these sections are prone to wave action, washing away fine material and leaving coarse fractions. At the beach, eolian transport may similarly contribute to this with a net loss of fine fractions to the dunes which, at least when being vegetated, act as sinks.

As can be seen from **Figure 5**, the excavated material, taken from sedimentation areas, is more sorted than the material left at the exposed erosion areas. However, onset of erosion takes place immediately and within 1 year sorting grade of the wave-action exposed sites fall back to the initial value. A similar trend can be observed at the beach, but readjustment is not finished within a year probably because eolian transport being a less efficient erosion process compared to continuous wave action.

All data presented so far indicate fast erosion of fine material at the beach and the offshore water sections. Originating from sedimentation sinks, the excavated material is rich in organic material (see **Table 1**) and, consequently, nutrients. This can be seen by the high concentrations of dissolved nitrogen species and phosphate in the pore water of the fresh material shown in **Figure 6**. Exposed to erosion, these nutrients are released into the water column, resulting in a fast drop of pore water concentrations which fell back to the original levels within half a year.

Grain size/ mm	Sorting grade	Water content/% of wet weight	Organic content/% of dry weight	Carbonate content/% of dry weight
0,270 ± 0,005	1388 ± 0,460	8658 ± 1248	10,090 ± 1450	0,501 ± 0,060

Grain size in mm, sorting grade, water content in % of total wet weight, organic and carbonate content in % of total dry weight are shown, both mean and standard deviation.

Table 1.
Sediment characteristics of the nourished sediment taken from sediment cores at the sand depot site.

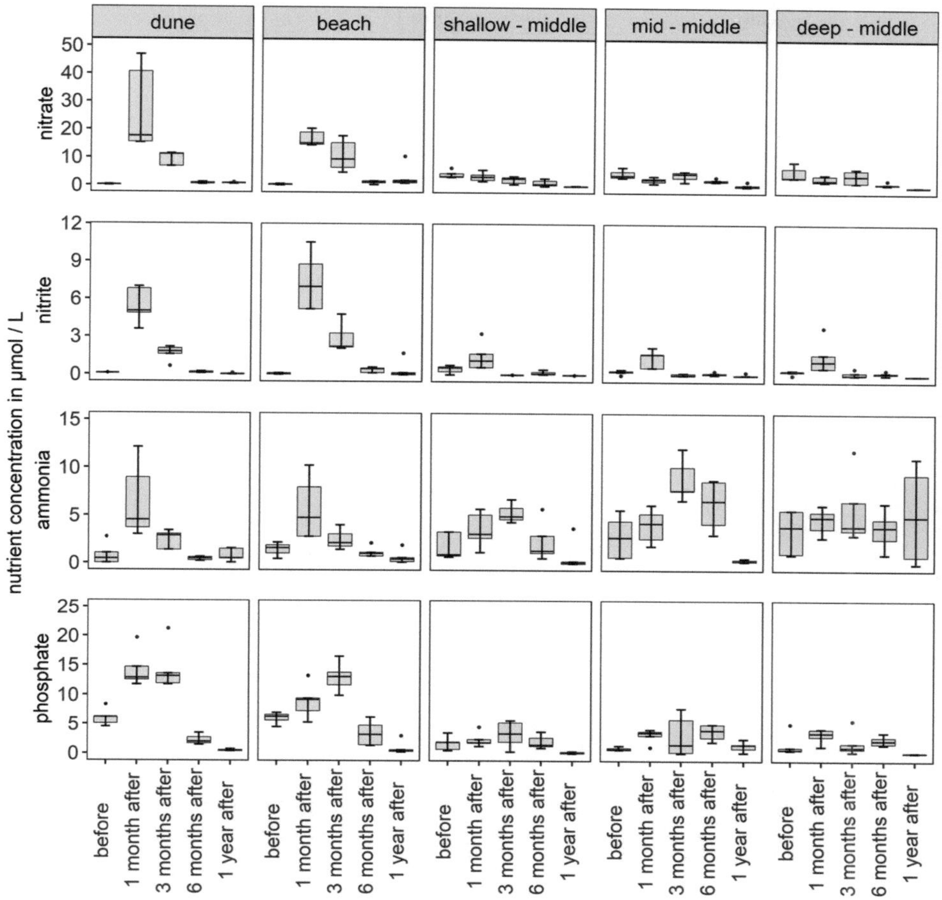

Figure 6.
Temporal development of the nutrient concentrations (nitrate, nitrite, ammonia, and phosphate) in µmol per L in the pore water of the sediment cores taken at the nourishment site (Ahrenshoop): Ahrenshoop at the dune, the beach, and the midsections between groyne lines in front of the beach ("shallow"), halfway the groyne line ("mid"), and at the end of the groyne line ("deep"). At the nourishment site, samples were taken before and during the nourishment as well as after the nourishment: 1 month, 3 months, 6 months, and 1 year after.

In the water column, the effect is even less pronounced. Phosphate and reduced nitrogen species exhibited elevated levels only during nourishment, when being washed out directly (**Figure** 7). After nourishment, the remaining release from pore water (compare **Figure 6**) seems to be converted immediately by plankton organisms which have been shown to be incredibly fast in nutrient uptake, for example, incorporating pulses of DIP into BM within 30 minutes [37]. This does not apply for nitrate, where elevated levels could be observed for about half a year. In this case, seasonality of the nutrient limitation regime must be taken into account. As shown by several authors [38, 39], autotrophs of the coastal systems of this part of the Baltic are N-limited during summer only, whereas in winter, energy limitation is dominating. Decrease of DIN therefore occurs at late spring, when N-limitation becomes effective and priming cyanobacterial development [40]. Nitrite and ammonia are not only N-sources for autotrophs but also used by prokaryotes as a source for energy as long as oxygen is available [41]. Therefore, the sharp drop of these oxygenation-sensitive

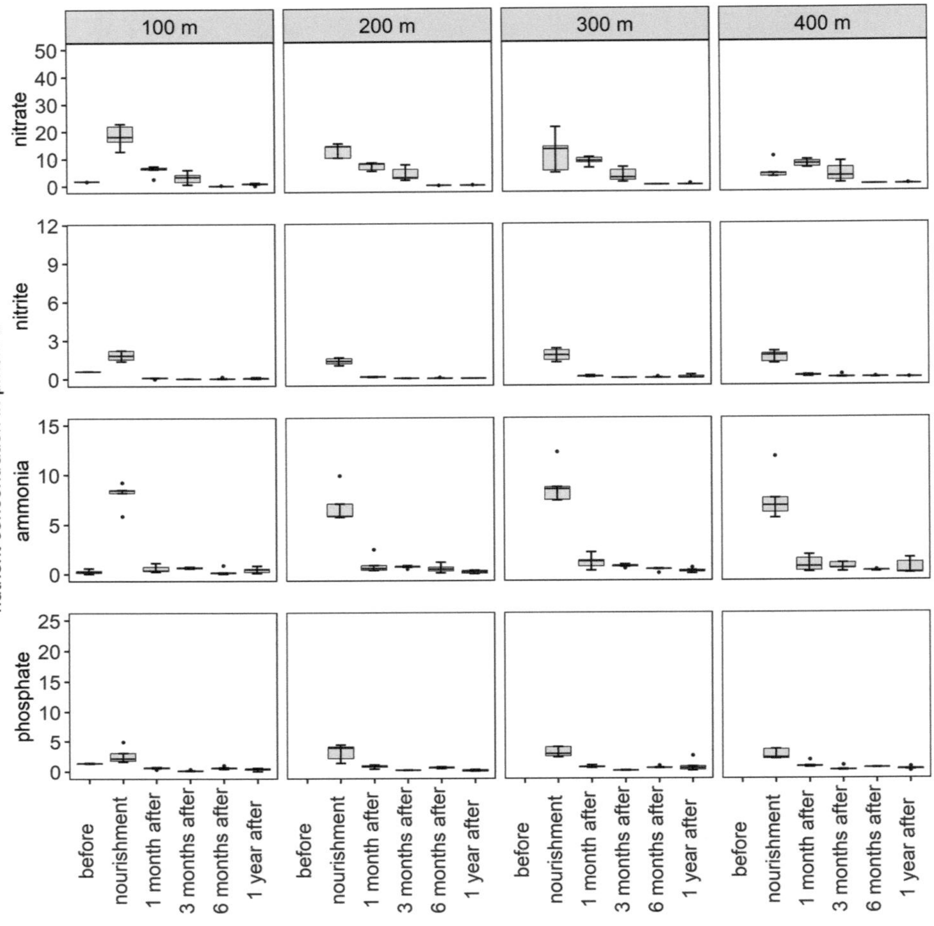

Figure 7.
Temporal development of the nutrient concentrations (nitrate, nitrite, ammonia, and phosphate) in µmol per L in the surface water samples taken at the nourishment site (Ahrenshoop) at different distances to the shore: 100, 200, 300, and 400 m. At the nourishment site, samples were taken before and during the nourishment as well as after the nourishment: 1 month, 3 months, 6 months, and 1 year after.

N-species reflects bacterial activity, converting them to nitrate and/or N2 [41]. However, bacterial growth requires phosphate, and the increased bacterial activity by organic material and reduced N-species provided by the nourishment here contributes to the fast decline in DIP observed.

The concentration of suspended solids (TSS) in the water column increased within a range of minimum 100 to 400 m from the nourished shore during the nourishment process. The resulting turbidity plume, as illustrated in **Figure 8**, extended along the coast for a minimum of 4 km. It is noteworthy that the sand nourishment occurred sequentially in specific locations, estimated to span approximately 50 to 100 m in length, with no simultaneous actions taking place. As soon as one of those locations was finished, the nourishment moved to the next section. Despite this, the turbidity originating from where the sand was nourished, was carried by waves and currents throughout the surrounding water column, dispersing along the coast. The precise dimensions of this turbidity plume were not determined due to unexpected extension and limitations in the sampling plan, which allowed for sampling within a 4 km

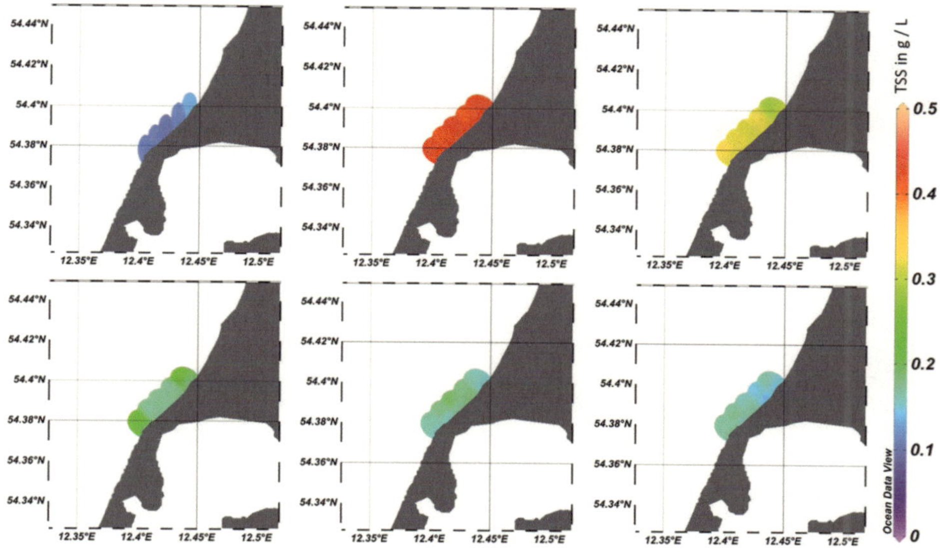

Figure 8.
Mean total suspended solids (TSS) in g per L in the surface water samples at the nourishment site (Ahrenshoop), from left to right: Upper row – Before, during, and 1 month after the sand nourishment; lower row – 3 months, 6 months, and 1 year after the nourishment. Red colors indicate high TSS concentrations, while blue colors indicate a lower concentration. The samples were taken along a 4 km long coastal section and up to 400 m away from the beach.

radius along the coast of Ahrenshoop. But in this radius of 4 km, elevated turbidity was measured during the sand nourishment. The extension of the turbidity plume into the Baltic Sea cannot be precisely determined, but we assume it to exceed 400 m, as we measured still elevated turbidity at this distance. The turbidity plume caused by sand nourishments are therefore not spatially limited to the exact nourishment site but can spread a long distance along the coast and into the open sea. This is especially important for coasts with a strong lateral current, such as the Baltic Sea.

Although elevated turbidity persisted 1 month post-nourishment, it gradually diminished and was no longer detectable by 6 months after the nourishment, see **Figure 9**. The organic content of these suspended solids was generally low and even lower post-nourishment, constituting only about 25% of the TSS. Homogeneity of TSS concentration throughout the water column during sand nourishments was demonstrated in Ahrenshoop [26] and by similar studies elsewhere [42].

In contrast to nourishments in other coastal areas, TSS levels were notably higher during the Ahrenshoop nourishment [42, 43]. This difference could be attributed to either a generally higher basal TSS in the Baltic Sea or seasonal variations. However, seasonal influence was ruled out as measurements taken 1 year post-nourishment occurred during the same season as the nourishment itself. One year after the coastal intervention, during the same seasons as the nourishment was performed, TSS levels returned to a low state, resembling pre-nourishment conditions. This suggests that the baseline turbidity of the Baltic Sea might be inherently higher than that of other oceans.

The sedimentation process was not quantified due to the low particle concentration, averaging around 0.4 g/L during the nourishment. Consequently, the heightened turbidity did not impact the submerged vegetation, primarily composed of

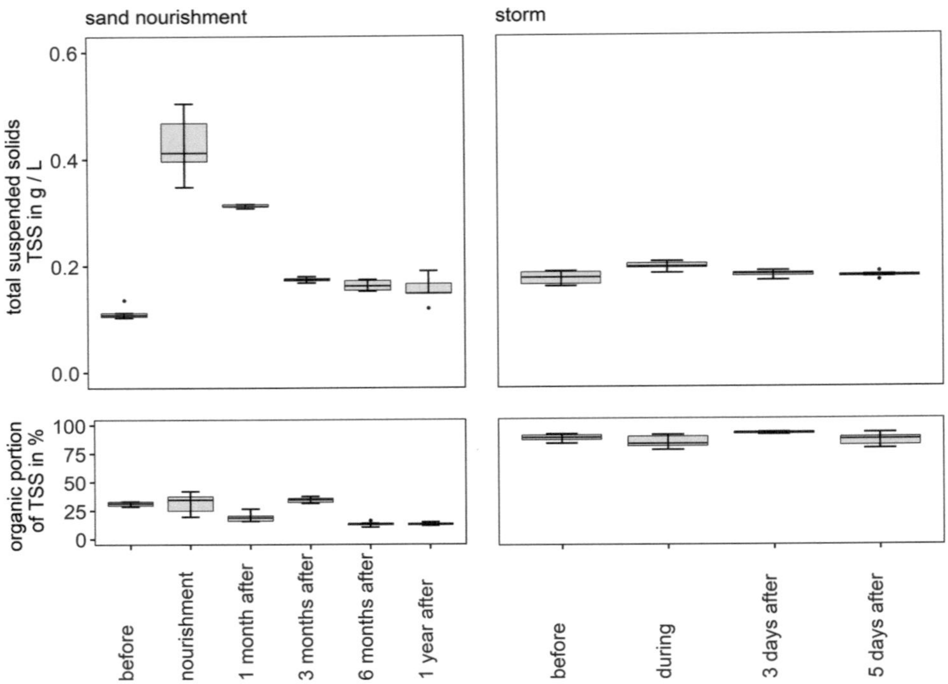

Figure 9.
Total suspended solids (TSS) in g per L and organic portion of TSS in % in the surface water samples at 100 m distance to shore at the nourishment site (Ahrenshoop, on the left side) in comparison with the TSS during a storm (Graal Mueritz, on the right side). At the nourishment site, samples were taken before and during the nourishment as well as after the nourishment: 1 month, 3 months, 6 months, and 1 year after. The storm samples were taken before and during the storm, as well as three and 5 days after the storm.

macroalgae attached to hard substrates. Despite sediment conditions indicating a potential habitat based on mean grain size, seagrass was not identified in the area. The macroalgae predominantly attached to wooden groynes near the shore, representing the sole hard substrate in that area.

To assess and classify such an artificial event and its repercussions, it was essential to compare it with a natural event, such as a storm event (**Figure 9**). During the sand nourishment, a significant amount of suspended material was observed in the water column, but the organic component was relatively low. Conversely, during the storm, the situation was reversed. Although the total suspended solids (TSS) concentration was lower than during the nourishment, the organic fraction increased. This pattern persisted in measurements taken before and after the storm. Overall, the turbidity caused by the storm diminished within a few days, showing pronounced differences with respect to longevity in comparison to the turbidity plume caused by sand nourishment.

The explanation for the different extension, both spatially and temporarily of those turbidity plumes, lies in the organic content of the sediment. At the extraction site, where sediment was sourced for nourishment, the organic content remained around 10% of the wet weight (**Table 1**). However, due to the extraction technique, these organic particles were washed out before reaching the nourishment site. After extracting the sediment, the drain of excess water takes place before the vessel moves to the nourishment site [42], creating another turbidity plume. This plume consists of very fine sediment fractions indicating that most of the organic content is left at the dredging site [44]. As a result, the sediment arriving at the nourishment site had low organic material, causing the turbidity plume to consist mainly of inorganic material like very fine sands and silts. The organic content at coastal sites where sand nourishments take place is usually decreased averaging below 2% [45]. Even months after the nourishment, the organic content in the suspended solids in the water column remained very low. In contrast, at the storm site, the sediment is richer in organic content as no organic-poor sediment is nourished there, leading to suspended solids composed mostly of organic material.

A substantial portion of the nourished sediment comprised very fine grain sizes (**Figure 10**). Approximately 10% of the sample fell below 63 μm, and nearly 20% ranged around 100 μm. The predominant fraction, constituting over 60% of the sediment, fell between 200 and 500 μm, categorized as middle sand. Coarser fractions collectively accounted for less than 10%. These findings align with the composition of the turbidity plume at the nourishment site, mainly consisting of inorganic but very fine material. This material was rinsed from the beach during the nourishment, accumulating in the water column and becoming suspended from the sea floor by waves and currents.

In addition to an elevated concentration of total suspended solids (TSS) during storm events, there was also an increase in nutrient concentrations (**Figure 11**). However, this increase was relatively small compared to the impact of the sand nourishment. Nitrite and phosphate, in particular, exhibited minimal changes, while nitrate and ammonia concentrations mostly increased, returning to pre-storm levels after 5 days. The nutrient concentrations during the storm were lower than during the nourishment because no additional sediment from deeper layers of the sea bottom, enriched with nutrients, was deposited at the storm site. Consequently, there were no heightened nutrient concentrations within the sediment that could be washed out by increased waves and currents during the storm event.

Regrettably, all samples collected during the storm event were taken from the water surface due to the impossibility of sampling from deeper layers. The paramount concern for human safety during the samplings led to challenging conditions, hindering sampling from multiple water depths. The waves and currents posed risks,

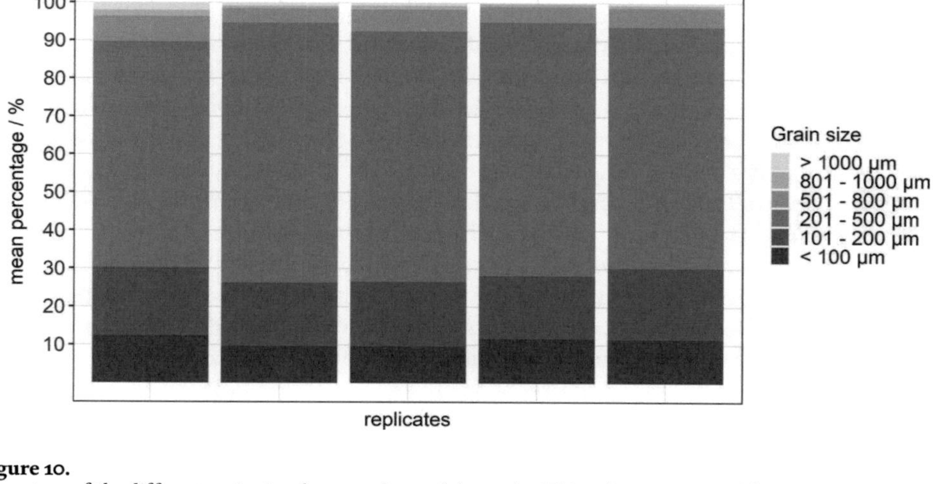

Figure 10.
Percentage of the different grain size classes at the sand depot site. This sediment was used for the sand nourishment in Ahrenshoop.

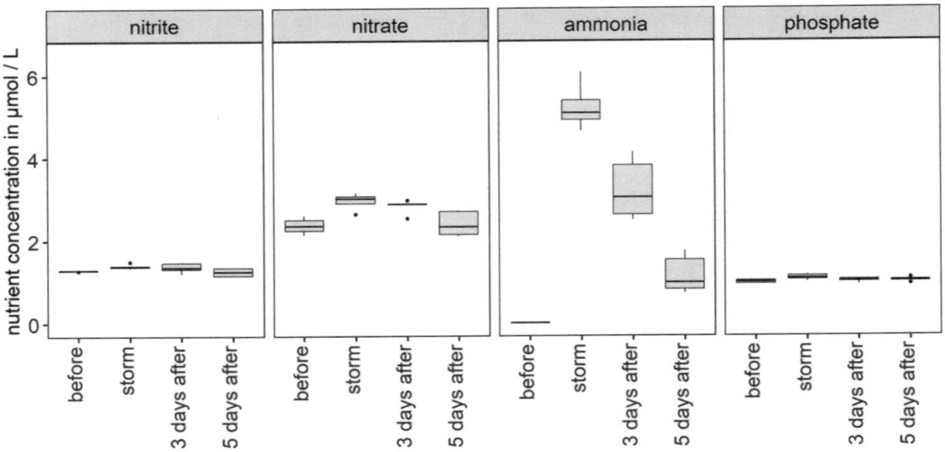

Figure 11.
Temporal development of the nutrient concentrations (nitrate, nitrite, ammonia, and phosphate) in µmol per L in the surface water samples taken at the storm site (Graal Mueritz) before, during, and after the storm.

making secure sampling unfeasible. As a result, it is plausible that TSS and nutrient concentrations during storms may differ between surface samples and samples from the sea bottom, with the latter potentially exceeding concentrations measured before.

4. Critical interpretation and inter-comparison

Increased turbidity is a common observation during and after sand nourishments in various studies [9, 46–48]. But as the baseline turbidity in the Baltic Sea, ranging usually between 0.1 and 0.3 g/L, already surpasses that of other oceans by the factor 50 [29], and this results in significantly higher turbidity levels during and post-sand nourishments. This regional irregularity underscores the unique environmental dynamics influencing the impact of coastal activities on turbidity in the Baltic Sea.

The results of this research indicate that the effects of sand nourishments disappear within an acceptable period of about 6 months at the deposition site, but also indicate strong effects on the decomposition site. Only limited studies focus on the sediment deposition site, the source of sediment for sand nourishment. The existing research highlights that habitat restoration may require decades once sediment is extracted and depending on the extraction method, large holes in the seafloor can form, filling with organic material and leading to anoxic conditions. Consistent with these findings and shown here, a substantial part of organic material and fine grain size classes are lost already during the extraction process. This means that buried organic material is resuspended in the water column and subsequently forms a fluffy layer, which alters the oxygen balance of the upper sediment layers and has a negative effect on the infauna at the extraction site [49]. Consequently, the turbidity plume and nutrient concentrations at the extraction site should be investigated and taken into account as well.

Most of these effects are not in the focus of the local stakeholders responsible for coastal protection, but should be assessed in more detail to allow for optimization of the whole coastal process.

It is also important to note that this study, comparing the impact of nourishments and storms on sandy shores in the Baltic Sea, warrants careful consideration.

DOI: http://dx.doi.org/10.5772/intechopen.1004905

Variations in TSS concentrations between surface and sea bottom samples are possible, but collection of the latter was not possible while prioritizing human safety during sampling. Therefore, further research should include turbidity measurements at multiple water depths during storm events to rule out the possibility of stratification of turbidity even at low water depths, as is the case in the Baltic Sea. A comparison at calm weather conditions was already performed [29]. Ongoing advancements in measuring devices will help to enhance data accuracy, as new devices are currently in development and construction.

5. Conclusion and assessment

The study assessed the effects of beach nourishment with respect to the impact on grain size distribution and nutrient concentration changes at the aquatic and terrestrial part. It must be taken into account that the studied sites are located in the microtidal Baltic Sea, differing largely from tidal coasts with respect to wave action along the littoral coastal gradient.

The results showed that the sand nourishment resulted in a long-lasting shift of grain size distribution and sorting grade of dune systems and beaches due to the increase of finer sediments, whereas in the aquatic part, the impact vanished within about 6 months. For nutrients, pore water concentrations of dissolved inorganic nitrogen as well as phosphate fell back to the original values within 6 months; in the open water, the effect of the nourishment campaign vanished even faster except for nitrate. Elevated nitrate concentrations were still detectable 3 months after the sand nourishment. This difference between nutrient kinetics can be explained by the seasonal limitation regime of the study area with an onset of nitrogen limitation in early summer.

For turbidity, the same applies. After 1 year, TSS returned from 0.4231 ± 0.0110 g/L during the nourishment to levels similar to previous to the nourishment with 0.1634 ± 0.0211 g/L, which is in good agreement with the loss of fine material in the shallow water regions of the nourishment area. Summarizing sand nourishment can be characterized as a heavy, but surprisingly short-term impact for the aquatic part of the coastal ecosystem as most effects can no longer be detected after 6 months. For the terrestrial part, mid-term effects at least must be taken into account lasting for at least 6 months, most prominent changes in grain size distribution of the dune influencing its eolian erosion resistance and consequently also alongshore sand transport.

Comparing sand nourishment with a storm event, striking differences were found especially with respect to nutrients. In general, the nutrient concentration was lower with around 5.9419 ± 1.5221 µmol/L during storm events than during the nourishment, where concentrations reached 7.8270 ± 1.8676 µmol/L. Ammonia exhibited the most pronounced kinetics after a storm, which can be explained by the high organic content of the resuspended material. For quantification of the global effects, it should be taken into account that a storm event impacts large areas, whereas sand nourishment concentrates on small spots. However, in both cases, the nutrient pulses were taken up by the biota within short time.

While concluding that the nourishment effects are vanishing within an acceptable time frame of about 6 months at the deposition site, the results also point to heavy impact on the extraction site. But another problem still to be investigated is toxic substances. During extraction, continuous control for explosives and military debris takes place. But other toxic substances have polluted the coastal seas as well, part of them accumulating in sedimentation sinks [50]. Routine assessment for such substances should be performed to prevent the pollution of recreation areas via sand nourishment.

The study presented here must be taken as a site-specific example where the pre-nourishment situation should not be mistaken as a "natural state." At Ahrenshoop, the neighboring cliff, which under natural conditions would be the source of material to keep the beach's matter balance, has been protected by a wall. Consequently, natural beach erosion is not balanced anymore by cliff erosion with consequences also for the factors (pore water nutrient content and grain size distribution) studied here. The pre-nourishment data therefore do not reflect a natural situation, but the equilibrium state of an altered system is kept by repeated sand nourishments. For assessing a natural state, more research on nutrient and sediment dynamics on unaltered cliff erosion-beach deposition systems are required.

Acknowledgements

This work was supported by a BMBF-funded project (ECAS-Baltic, FKZ 03F0860F) which is gratefully acknowledged. Special thanks go to the crew of the LIMANDA, the StALU MM, and all scientific divers supporting the samplings.

Author details

Daniela Glueck* and Hendrik Schubert
University of Rostock, Biosciences, Aquatic Ecology, Rostock, Germany

*Address all correspondence to: daniela.glueck@uni-rostock.de

References

[1] Firth LB, Knights AM, Bridger D, Evans AJ, Mieszkowska N, Moore P, et al. Ocean sprawl: Challenges and opportunities for biodiversity management in a changing world. In: Hughes RN, Hughes DJ, Smith IP, Dale AC, editors. Oceanography and Marine Biology: An Annual Review. Vol. 54. Boca Raton: Taylor & Francis; 2016. pp. 189-262

[2] Dugan JE, Airoldi L, Chapman MG, Walker SJ, Schlacher T. Estuarine and Coastal Structures. In: Treatise on Estuarine and Coastal Science. Amsterdam: Elsevier; 2011. pp. 17-41

[3] Schoonees T, Gijón Mancheño A, Scheres B, Bouma TJ, Silva R, Schlurmann T, et al. Hard structures for coastal protection, towards greener designs. Estuaries and Coasts. 2019;**42**(7):1709-1729

[4] van Rijn LC. Design of hard coastal structures against erosion; Available from: www.leovanrijn-sediment.com [Accessed: October 17, 2023]

[5] Beck M, Airoldi L. Loss, status and trends for coastal marine habitats of Europe. In: Gibson R, Atkinson R, Gordon J, editors. Oceanography and Marine Biology: An Annual Review. Vol. 45. Boca Raton: Taylor & Francis; 2007. pp. 345-405

[6] Pontee N, Narayan S, Beck MW, Hosking AH. Nature-based solutions: Lessons from around the world. Proceedings of the Institution of Civil Engineers - Maritime Engineering. 2016;**169**(1):29-36

[7] Silva R, Martínez ML, Hesp PA, Catalan P, Osorio AF, Martell R, et al. Present and future challenges of coastal erosion in Latin America. Journal of Coastal Research. 2014;**71**:1-16

[8] Kortekaas S, Bagdanaviciute I, Gyssels P, Huerta JMA, Héquette A. Assessment of the effects of marine aggregate extraction on the coastline: An example from the German Baltic Sea Coast. Journal of Coastal Research. 2010;**51**:205-214

[9] Nordstrom KF, Jackson NL, Pranzini E. Beach sediment alteration by natural processes and human actions: Elba Island, Italy. Annals of the Association of American Geographers. 2004;**94**(4):794-806

[10] Speybroeck J, Bonte D, Courtens W, Gheskiere T, Grootaert P, Maelfait J-P, et al. Beach nourishment: An ecologically sound coastal defence alternative? A review. Aquatic Conservation: Marine and Freshwater Ecosystems. 2006;**16**(4):419-435

[11] Nordstrom KF. Beach nourishment and coastal habitats: Research needs to improve compatibility. Restoration Ecology. 2005;**13**(1):215-222

[12] Grunewald R. Assessment of damages from recreational activities on coastal dunes of the southern Baltic Sea. Journal of Coastal Research. 2006;**225**:1145-1157

[13] Grunewald R, Schubert H. The definition of a new plant diversity index H′dune for assessing human damage on coastal dunes - derived from the Shannon index of entropy H′. Ecological Indicators. 2007;**7**(1):1-21

[14] Jordan P, Fröhle P. Bridging the gap between coastal engineering and nature conservation? Journal of Coastal Conservation. 2022;**26**(2):4

[15] Tiede J, Jordan C, Moghimi A, Schlurmann T. Long-term shoreline changes at large spatial scales at the Baltic Sea: Remote-sensing based assessment and potential drivers. Frontiers in Marine Science. 2023;**10**:1207524

[16] Miyanishi K, Johnson EA. Coastal dune succession and the reality of dune processes. In: Plant Disturbance Ecology. Amsterdam: Elsevier; 2021. pp. 253-290

[17] Arens SM. Patterns of sand transport on vegetated foredunes. Geomorphology. 1996;**17**(4):339-350

[18] Krause JC, Diesing M, Arlt G. The physical and biological impact of sand extraction: A case study of the Western Baltic Sea. Journal of Coastal Research. 2010;**51**:215-226

[19] Mielck F, Hass HC, Michaelis R, Sander L, Papenmeier S, Wiltshire KH. Morphological changes due to marine aggregate extraction for beach nourishment in the German bight (SE North Sea). Geo-Marine Letters. 2019;**39**(1):47-58

[20] Szymelfenig M, Kotwicki L, Graca B. Benthic re-colonization in post-dredging pits in the Puck Bay (Southern Baltic Sea). Estuarine, Coastal and Shelf Science. 2006;**68**(3-4):489-498

[21] Uścinowicz S, Jegliński W, Miotk-Szpiganowicz G, Nowak J, Pączek U, Przezdziecki P, et al. Impact of sand extraction from the bottom of the southern Baltic Sea on the relief and sediments of the seabed. Oceanologia. 2014;**56**(4):857-880

[22] Staudt F. Sandaufspülungen für den Küstenschutz. Hamburg: Das Vorhaben STENCIL; 2019

[23] Staudt F, Gijsman R, Ganal C, Mielck F, Wolbring J, Hass HC, et al. The sustainability of beach nourishments: A review of nourishment and environmental monitoring practice. Journal of Coastal Conservation. 2021;**25**(2):34

[24] Jarmalavičius D, Šmatas V, Stankūnavičius G, Pupienis D, Žilinskas G. Factors controlling coastal erosion during storm events. Journal of Coastal Research. 2016;**75**(sp1):1112-1116

[25] Kullenberg G. Chapter 3: Physical oceanography. In: Voipio A, editor. The Baltic Sea. Amsterdam: Elsevier Scientific Publishing Company; 1981. pp. 135-181

[26] Winterhalter B, Flodén T, Ignatius H, Axberg S, Niemisto L. Chapter 1: Geology of the Baltic Sea. In: Voipio A, editor. The Baltic Sea. Amsterdam: Elsevier Scientific Publishing Company; 1981. pp. 27-121

[27] Staatliches Amt für Landwirtschaft und Umwelt Mittleres Mecklenburg. Regelwerk Küstenschutz Mecklenburg-Vorpommern; Available from: https://www.stalu-mv.de/mm/Themen/K%C3%BCstenschutz/Regelwerk-K%C3%BCstenschutz-Mecklenburg%E2%80%93Vorpommern/ [Accessed: November 30, 2023]

[28] Tiepolt L, Pöhlmann B, Behl H. Ausführungsplanung zum Vorhaben Dünenverstärkung Ahrenshoop - Vordarß. Rostock; 2021

[29] Glueck D. Comparison of high turbidity events: Sand nourishments and storm events on sandy beaches at the Baltic Sea, Germany. Marine Pollution Bulletin. 2023;**194**(Pt A):115389

[30] Sindowski K-H. Sedimentpetrographische Methoden zur Untersuchung sandiger Sedimente 1938;**29**(3-5):196-200

[31] Schlungbaum G. Untersuchungen über die Sedimentqualität in den Gewässern der Darß-Zingster-Boddenkette unter Berücksichtigung der Stoffaustauschprozesse zwischen Wasser und Sediment. Habilitation thesis, University of Rostock; 1979

[32] Czensny R. Wasser-, Abwasser- und Fischereichemie. Leipzig: Deutscher Verlag für Grundstoffindustrie; 1961

[33] Grasshoff K, Kremling K, Ehrhardt M. Methods of Seawater Analysis. 3rd ed. Weinheim, Germany: Wiley VCH Verlag GmbH; 1999

[34] Eichmanns C, Schüttrumpf H. Investigating changes in Aeolian sediment transport at coastal dunes and sand trapping fences: A field study on the German coast. JMSE. 2020;**8**(12):1012

[35] Anderson RS. Erosion profiles due to particles entrained by wind: Application of an eolian sediment-transport model. Geological Society of America Bulletin. 1986;**97**(10):1270

[36] van Rijn LC. Coastal erosion and control. Ocean & Coastal Management. 2011;**54**(12):867-887

[37] Schumann R, Berthold M, Eggert A, Blindow I, Forster S, Schubert H. Short-term variability, long-term trends and seasonal aspects in the Darß-Zingst Bodden chain. In: Schubert H, Müller F, editors. Southern Baltic Coastal Systems Analysis. Vol. 246. Cham: Springer International Publishing; 2023. pp. 117-128

[38] Schiewer U. Darß-Zingst Boddens, northern Rügener Boddens and Schlei. Ecological Studies. 2008;**197**:35-86

[39] Schubert H, Wasmund N. Das Phytoplankton des Strelasundes und des Kubitzer Boddens. Meer und Museum. 2005;**18**:83-89

[40] Laamanen M, Kuosa H. Annual variability of biomass and heterocysts of the N_2-fixing cyanobacterium *Aphanizomenon flos-aquae* in the Baltic Sea with reference to *anabaena* spp. and *Nodularia spumigena*. Boreal Environment Research. 2005;**10**:19-30

[41] Schneider B, Dellwig O, Kuliński K, Omstedt A, Pollehne F, Rehder G, et al. Biogeochemical cycles. In: Snoeijs-Leijonmalm P, Schubert H, Radziejewska T, editors. Biological Oceanography of the Baltic Sea. Dordrecht: Springer Netherlands; 2017. pp. 87-122

[42] Capello M, Cutroneo L, Ferranti MP, Budillon G, Bertolotto RM, Ciappa A, et al. Simulations of dredged sediment spreading on a *Posidonia oceanica* meadow off the Ligurian coast, Northwestern Mediterranean. Marine Pollution Bulletin. 2014;**79**(1-2):196-204

[43] Wilber DH, Clarke DG, Burlas MH. Suspended sediment concentrations associated with a beach nourishment project on the Northern coast of New Jersey. Journal of Coastal Research. 2006;**225**:1035-1042

[44] Qui W, Sarkar S. Marine technology and SNAME news. Society of Naval Architects & Marine Engineers. 2006;**43**

[45] Colosio F, Abbiati M, Airoldi L. Effects of beach nourishment on sediments and benthic assemblages. Marine Pollution Bulletin. 2007;**54**(8):1197-1206

[46] Manning LM, Peterson CH, Bishop MJ. Dominant macrobenthic populations experience sustained impacts from annual disposal of fine sediments on sandy beaches. Marine Ecology Progress Series. 2014;**508**:1-15

[47] Roegner GC, Fields SA, Henkel SK. Benthic video landers reveal impacts of dredged sediment deposition events on mobile epifauna are acute but transitory. Journal of Experimental Marine Biology and Ecology. 2021;**538**:151526

[48] Roman-Sierra J, Navarro M, Muñoz-Perez JJ, Gomez-Pina G. Turbidity and other effects resulting from Trafalgar sandbank dredging and Palmar Beach nourishment. J. Waterway, port, coastal. Ocean Engineering. 2011;**137**(6):332-343

[49] Zabarte-Maeztu I, Matheson FE, Manley-Harris M, Davies-Colley RJ, Hawes I. Fine sediment effects on seagrasses: A global review, quantitative synthesis and multi-stressor model. Marine Environmental Research. 2021;**171**:105480

[50] Lehtonen KK, Bignert A, Bradschaw C, Broeg K, Schiedek D. Chemical pollution and ecotoxicology. In: Snoeijs-Leijonmalm P, Schubert H, Radziejewska T, editors. Biological Oceanography of the Baltic Sea. Dordrecht: Springer Netherlands; 2017. pp. 547-587

Chapter 8

Sediment Profiles and Recording the Effects of Anthropogenic Activities

Denise Balestrero Menezes, Daniele Toyama and Clara Bianchi Bahia

Abstract

The sediments of streams and rivers partly preserve the records of the activities that surround them. In natural areas, mineral sediments are deposited and organic plant material is carried by rainfall. If in urban or densely cultivated areas, the records are complemented by other organic materials, and chemicals and solid particles from human activities. Studying sediment profiles of small watercourses throughout their hydrological cycle allows us to observe oscillations in sediment inputs and correlate them with the surrounding uses. This makes it possible to observe influences on erosion processes in areas of exposed soil due to agriculture, on the reduction of water flow in primary channels due to soil sealing and on changes in the solid content of sediments due to the incorporation of anthropogenic materials, like microplastics. The methodological procedures presented in this chapter and used in two case studies carried out the obtention of sediment profiles ranging from 12 to 71 cm in length, enabling in-depth analysis of the content of anthropogenic particles. In the urban basin studied, we found up to 5160 part./kg of microplastics, and in the rural basin, there were only a few sparse fragments, highlighting urban waste and activities as the main source of these materials.

Keywords: sampling, strata, core sampler, microplastics, chemical pollution, flow reduction

1. Introduction

In order to assess the state of integrity of soils, waters and sediments in urban water bodies, it is necessary to characterise the quality of these elements of the physical environment. Many of the interactions between elements occur through water flows from springs and streams that drain urban and rural catchments.

The influence of human beings as modifying agents of the physical environment has long been pointed out. The extent of these modifications and the way they affect the physical and chemical properties of soils, sediments and waters depends on the type of anthropogenic action and the intensity of the alteration caused. The diversity of chemical products and materials has never been greater than it is today [1].

Water resources and their quality are closely related to the characteristics of the geoenvironment (geological formations, soils, surface and underground water bodies, gases and organisms), which give them geogenic signatures; anthropogenic activities, through modifications to geological, physical, chemical and biochemical processes, modify these signatures, incorporating elements that make it possible to identify them [2].

Rivers are the main carriers of sediment loads. Water and wind carry the soil particles that break off to the watercourses, which transport them to the lower reaches. Human action on the vegetation cover, exposing the soil and causing rivers to silt up, can aggravate the natural surface processes. Changes in soil cover allow more rainwater to run off the surface and carry particles of various natural and artificial products into watercourses, increasing sedimentation and changing their compositional characteristics [3].

The study of sediment in watercourses and its composition provides an assessment of the natural components originating from the weathering and erosion of the surrounding rocks and soils, as well as materials originating from human industrial, urban and rural activities upstream, such as plastic, metals, wood, paper, cardboard, bricks and other construction waste that can be incorporated into river sediment. The composition reflects greater or lesser anthropisation; one of the indicators is the presence of plastics (macro or micro), whose dispersion and effects on terrestrial aquatic environments are still poorly studied. Much of the plastic produced is a source of pollution for the geoenvironment and in terrestrial water bodies; it accumulates in places with low water flow velocity, such as lakes, floodplains and meanders [4–6].

Farming activities are also a major cause of environmental change, interfering with water dynamics as they alter surface geological materials through the revolving of materials excavated, turned over and landfilled, modifying the natural structure, destroying the layer of organic matter and allowing the leaching of materials [7, 8].

These surface alterations have an impact on watercourses and transported sediment, leaving a record of variations in composition and volume in the deposits [9]. In addition, this process of sedimentary deposit formation can change over time due to variations in water flows, influencing the records found [10].

The content and characteristics of the sediment deposited can indicate an input of sediment greater than the transport capacity of the water flow (siltation), the flow of organic matter transported by torrents or the flow of anthropogenic materials or contaminants [11].

Another effect of anthropogenic alterations is that, during rainfall, a greater flow of water reaches the channel, with less infiltration. This suddenly concentrated flow generates marginal erosion and the consequent loss of riparian vegetation. Legally protected, riparian vegetation, when denser and intact, has the capacity to retain part of the flow of sediment and other materials that can reach the watercourse, as well as retain erosion processes on the banks of watercourses.

Evidence of degradation of the physical environment makes it possible to assess the health of the water environment for aquatic fauna, for example. Moreover, they can guide the environmental recovery of the areas drained by the river basin.

In the history of humankind, the Neolithic Revolution represents a milestone in the development of the species and its capacity as modifiers of the physical environment. Previously, natural agents such as volcanic processes, erosion, landslides, weathering and the transport of materials were responsible for remodelling the surface and creating new deposits. After the Neolithic Revolution, there were several technological revolutions, and now humans, as agents modifying the environment, carry out excavations and earthworks, build cities, extract minerals and produce solid waste. It is important to emphasise the intensity of human actions on Earth, especially in comparison with the planet's lifespan.

As pointed out by [12], human activities in the 21st century move far more sediment per year than all the rivers on Earth, as well as convert half of all habitable land to agriculture. Moreover, rivers, the source of water necessary for life, are turned degraded, polluted and modified.

The human population, through its technology, is capable of irreversible negative changes and impacts. In this sense, the Anthropocene Working Group (AWG) studies whether human impacts can be so significant as to indicate a new geological epoch [13, 14]. Regardless of these chronostratigraphic definitions, in this post-World War II period, the Great Acceleration of population growth, industrialisation and use of mineral and energy resources in the middle of the 20th century, such as the extraction and processing of oil, producing fossil fuels and other derivatives, generated new persistent materials, among them synthetic polymers [15].

Plastics and their incorporation into our daily lives are well known. Their effects on the environment are still being studied.

Disposable plastic appeared in 1909 because of US legislation banning the use of communal cups on trains in order to prevent the spread of disease; in 1938, polyamide (fibres that make up fabrics) was invented, introducing plastic into the fashion industry. In the 1950s, fabrics made from polyester took over the market because they were cheap, were easy to wash and did not require ironing. Between 1970 and 1980, plastic began to be used in the manufacture of children's toys, spacecraft components and objects related to music and photography. In the 1990s, plastic was used in the automotive industry because it was a lightweight material, increasing the fuel efficiency of vehicles. In the 2000s, on the other hand, the call for environmental protection increased and the issue of recovering and recycling plastics gained attention [16].

In the drinks market, the first tests with plastic packaging took place in 1970 by Coca-Cola and went into commercial production in 1976 by Amoco for Pepsi Cola [17].

In Brazil, the use of plastic bags began in 1980, due to the rising cost of paper [18].

Of all the plastic produced in the world in 2015, 2.5 x 109 tonnes (30.12%) are in use, 0.6 x 109 tonnes (7.22%) were recycled, 0.8 x 109 tonnes (9.63%) were incinerated and the remaining 4.9 x 109 tonnes (59.03%) were disposed of in open dumps and landfills and dispersed in inappropriate places [5]. Much of this material will accumulate and sediment in water bodies.

Given its persistence in the environment, plastic is one of the most important constituents of future technofossils, since it is widely exploited by today's industry, is present on a large scale in the daily lives of the world's population [19] and occurs in various sedimentation environments.

Plastic in the environment receives a classification according to its size: macroplastics are those larger than 5 mm, microplastics are smaller than 5 mm [20] and nanoplastics are tens of nanometres or less [5]. Macroplastics are waste that is visible to the naked eye, such as bags, bottles, toys, fishing nets and packaging. Microplastics are not easily visible to the naked eye, being microspheres, pellets and secondary fragments derived from macroplastics and fabric fibres and according to [5] microplastics have a considerable capacity to cause biological damage. The diffusion of nanoplastics possibly occurs to the same extent as the previous ones, but due to the lack of methodology for their detection, they are little studied.

Sediments are a result of hydrosedimentological dynamics and their anthropisation have relation to these characteristics, but also due to the shape and complexity of the different land uses in the surrounding area [21–23].

In order to understand the diffusion of non-natural particles, our research group is trying to assess the extent to which the sediments are impacted, whether they are composed of anthropogenic materials, whether they are little anthropised or whether they indicate alterations resulting from the environment.

The objectives of this chapter are:

- to present the methodology and procedures developed and adopted by the group, seeking to analyse anthropogenic changes and their recording in depth, which highlights changes in deposits over time;

- to present and discuss the results of two different applications.

To this purpose, we structured the text as follows: topic 2 describing the methodology and methods of analysis, topic 3 applying it to the determination of microplastics in sediments from urban and rural watercourses and topic 4 applying it to the identification of anthropisation of watercourses in rural areas, seeking to evaluate the different types of results with its application.

We end with a discussion of the results, comparison with other studies and conclusions.

2. Sampling and analysis methodology

The methodology for collecting sediment profiles and analysing composition and the presence of anthropogenic materials was developed looking for in-depth data, as these are still incipient. It was applied in two locations: first in an urban basin [24] and then in a rural basin [25], the results and specific aspects of which will be presented in sections 3 and 4.

We developed a corer-type sampler and used to collect the sediment profile, which tipe is indicate in studies that aim to verify changes on a temporal scale, since we can observe the successive sediment layers. The sampler consists of a two-inch, 1 m-long geomechanical PVC tube with an internal plunger for suction, working in a similar way to a syringe [26].

As these are shallow bodies of water, sampling takes place inside the channel by driving the sampler into the bed sediment while lifting the plunger, allowing the sample to enter. The sampler is then removed and the sample extracted by pushing the plunger.

In general, samples are taken in two separate collections over the course of a hydrological year in order to observe changes between periods of drought and periods of greater rainfall intensity, which are characteristic of the climate in the southeastern region of Brazil. In the case of climatic features with a different rainfall regime, collections during the year should be adapted.

The sample profiles collected are divided every 10 cm or when different layers occur (textural, compositional or colour variations) and stored.

The sub-samples are dried at room temperature and separated into particle size fractions (>4.75 mm, 4.75 to 1.18 mm and 1.18 mm to 75 μm and < 75 μm) by sieving using a sieve shaker. After sieving, the fractions are weighed and stored.

The particle size fractions referring to pebbles, gravel and other materials >1.18 mm are sorted with the naked eye or using a hand magnifier and tweezers. Anthropogenic fragments found are identified, separated, photographed and stored.

The fraction corresponding to sands between 1.18 mm and 75 μm is quartered using a metal spatula and glass base, prepared and analysed under a stereomicroscope. A quarter of the volume is separated for analysis and a third of the remaining material is kept for replication.

In an attempt to define the best preparation method, we carried out a comparative analysis of samples of this fraction [24] without degradation of organic matter (OM) and with degradation of organic matter. Degradation was chosen because it makes it easier to see the natural material compared to the artificial one, reduces analysis time and minimises possible errors arising from the similarity between plastic fibres and roots and other smaller fragments of organic matter.

The degradation of OM is carried out by adding 30% hydrogen peroxide (no filtering was carried out) to the sediment in beakers or porcelain crucibles [27, 28], using a sufficient volume to cover the sample, which is stirred with a glass rod and then covered to avoid contamination. After the reaction ceases, the samples are dried using a sand bath. Once drying is complete, H2O2 is added again, waiting for oxidation to finish and drying to take place.

At the end of the degradation process, the samples are analysed under a stereomicroscope, taking note of any unnatural particles present, recording them and identifying them where possible.

In the 2 studies presented below, due to their peculiarities, the materials were counted in different ways.

The <75 μm fraction can be sent for chemical analyses, which will not be presented in this chapter.

3. The Água Branca stream basin and plastics

The presence of plastics in the sediments was the focus of the methodology in this river basin in the interior of Brazil.

The aim of the study was to identify the occurrence of the artificialisation of river sediment and the deposition of anthropogenic material on the bed of water bodies in a river basin, correlating them with the characteristics of urban and rural use of the surrounding area.

The Água Branca Stream Basin drains a large part of the urban area of Itirapina, a municipality located in the interior of the state of São Paulo, Brazil, with an area of 564.603 km^2 and 16,148 inhabitants [29].

In this study, we selected sample points (shown in **Figure 1**) in streams located in areas of greater and lesser urbanisation in order to verify the sedimentation patterns of artificially produced materials.

The sampling campaigns took place in November 2018 (spring) and May 2019 (autumn). The first campaign occurred after a period of drought from April to July, followed by rainfall of approximately 50 mm until November. The second followed a period of heavier rainfall, reaching 180 mm in February. In May 2019, rainfall halved compared to the previous month.

We also observed the accumulation of man-made materials on the banks or in the bed of the stream. In these cases, we took photographs and collected specimens of interest.

This text will focus on the macro and micro scales of anthropogenic fragments.

Point, coordinates (WGS 84) and description	
P1 – 0210004; 7536916  Unnamed stream tributary of the right bank of the Água Branca stream. No riparian forest, close to the urban area and the railway line. Solid waste deposited on the banks. There is cardboard and plastic incorporated into the sediment.	P2 – 0210296; 7536552  Água Branca stream at the entrance to the urban area. This stretch has no riparian vegetation, with intense urban occupation on both banks. There is the presence of solid urban waste from the drainage system.
P3 – 0207176; 7537866  Água Branca stream, at a point located after the urban area and the destination of effluent from the Itirapina sewage treatment plant. Area with riparian vegetation.	P4 – 202587; 7566379  Água Branca stream, at a point located after the urban area and the reception of effluent from the Itirapina sewage treatment plant. Banks with riparian vegetation.
P5 – 0210358; 7536503  Água Branca stream in a rural area upstream of the urban area. The riparian forest complies with environmental legislation, but there is interference from an irrigation system for family agricultural production.	P6 – 0209396; 7536984  Água Branca stream in a rural area upstream of the urban area. The riparian forest complies with environmental legislation, but there is interference from an irrigation system for family farming.

Figure 1.
Description of the sediment collection points. Source: Author (2019).

The samples collected from the sediment profiles underwent the separation and sorting procedure presented in item 2. For more details on the specific procedures, read [24].

The material retained on the 4.75-mm and 1.18-mm sieves and observed with the naked eye, when there was evidence of plastic material, underwent a heat resistance test that allows plastic, organic matter and mineral grain identification. The description of this procedure is in [24].

The material in the fraction corresponding to sands from 1.18 mm to 75 μm had all the anthropogenic particles separated, photographed, identified (where possible), quantified and statistically evaluated [24].

Plastic, the most common artificial material today, is also present in the air. In this study, we considered the possibility of sample contamination throughout the sorting and sample preparation stage. Therefore, the samples from point P4 (Limoeiro Stream) were defined as blanks for this study. The fibres and other fragments observed at this point could be the effects of contamination and we disregarded them from the analyses, extrapolating to the other points.

Considering the geological substrate of the basin, composed of sandy sedimentary rocks, and the granulometric analysis, the sandy matrix sediments, with little silt and clay, correspond to what was expected. There is a predominance of quartz grains, some opaque minerals (some ferrous with magnetic attraction) and micas. We can find organic material incorporated into the sediment and artificial components such as plastic.

The main anthropogenic material observed was plastic (macro and microplastic). The economy of the study area have a not significant impact by industrial production, so the uses of the basin are predominantly urban (housing) and commercial (sale of finished products and provision of services); the artificial materials found in the samples are a reflection of this occupation. The following sections present the results and discussion of each point studied.

We observed the deposition of materials of artificial origin in the bed of water bodies at some points. The following subsections detail the points where we observed macroscopic traces of sediment artificialisation.

In the bed of the unnamed stream, in the urban area (P1), it is possible to see the deposition of artificial waste such as plastic and cardboard.

In the Água Branca Stream - urban area (P2), the collection occurred in the canalised section of the stream, where we found artificial material such as roof tiles, fragments of bricks, shoes and plastics deposited. We observed the presence of fish and aquatic macro invertebrates.

In the Tibiriçá Reservoir (P6), a place of transition from a lotic to a lentic environment, all the drainage from the upstream basin contributes to the composition of the sediment. The margin of the reservoir has an accumulation of artificial materials and in addition to the sediment pro-files we collected these deposited objects for characterisation.

Figure 2 shows some of these objects. The oldest container found dates back to 1993. This demonstrates the reservoir's capacity to retain material from the drainage network. All the specimens collected received solar radiation for years. All the items collected are perfectly preserved and the degradation observed in some items is due to mechanical abrasion (e.g. Axe deodorant packaging). The most recent packaging collected was dated 2015.

Although some of the material is undated, it is possible to infer its antiquity considering that it is packaging for products that are not in commercial circulation. The observations made show that artificially altered deposits occur mainly in environments with more intense sedimentation.

The sediment profiles collected ranged from 16 to 71 cm in length (longer in P6) and was possible divide them in 2 to 4 sub-samples.

Analysis of the sediment profiles showed that the finer fractions are the ones that make it possible to see the greatest compositional variability.

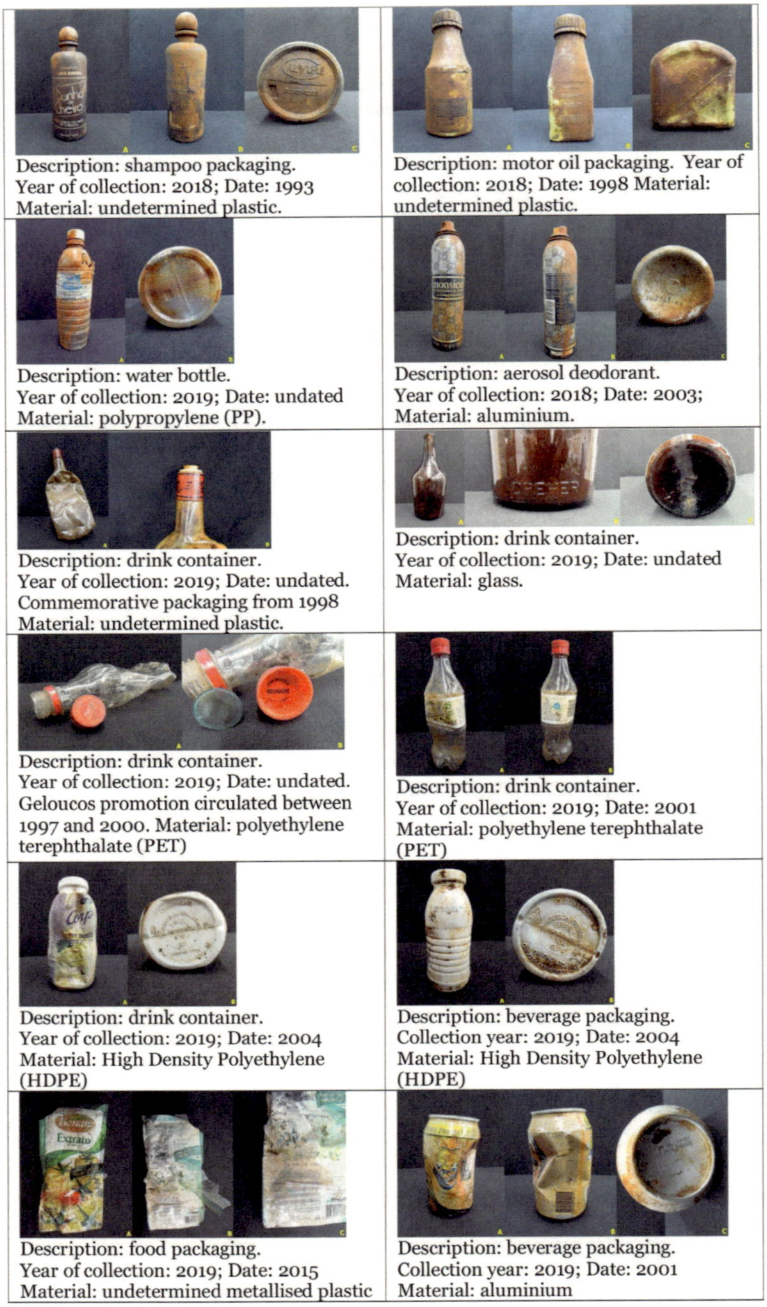

Figure 2.
Examples of objects collected at Tibiriçá Reservoir deposited within the sediment. Source: Author (2019).

In the macroscopic analysis we observed few fragments, consisting of microplastics, plant fibres and glass, with the exception of point P6, which showed a greater quantity of anthropogenic fragments at different sample depths. At point P5, in contrast to the samples from the other points where sand predominates, we observed a different sediment composition, with material similar to sawdust used to cover vegetable beds (organic matter).

In addition, we found sticks and leaves from the riparian forest that exists in this stretch. We observed organic matter (fragments of leaves, sticks, algae, macro invertebrate cocoons, seed shells) in the surface layer of point P4, while the deeper layers were mostly composed of sand. At this point, we did not observe the presence of artificial materials as expected.

Microscopic analysis revealed many fragments. **Table 1** shows the quantity of plastic fragments, plastic fibres and other materials (paper or glass) observed in the profiles studied (material between 1.18 mm and 75 μm in size).

Figure 3 shows some examples of the fragments recorded in 2018 and 2019.

A detailed analysis of the results follows in [24].

		Fragments		Fibres		Other		Total	
		2018	2019	2018	2019	2018	2019	2018	2019
P1.1	part./100 g	14	9	19	76	0	0	34	85
	part./kg	140	90	190	760	0	0	340	850
P1.2	part./100 g	0	49	22	130	0	0	22	179
	part./kg	0	490	220	1300	0	0	220	1790
P2.1	part./100 g	20	40	72	29	0	0	92	69
	part./kg	200	400	720	290	0	0	920	690
P2.2	part./100 g	8	8	65	40	4	0	77	48
	part./kg	80	80	650	400	40	0	770	480
P2.3	part./100 g	11	12	2	0	0	0	13	12
	part./kg	110	120	20	0	0	0	130	120
P3.1	part./100 g	2	1	13	16	0	0	15	17
	part./kg	20	10	130	160	0	0	150	170
P3.2	part./100 g	0	1	2	10	0	0	2	12
	part./kg	0	10	20	100	0	0	20	120
P3.3	part./100 g	0	3	0	0	0	0	0	3
	part./kg	0	30	0	0	0	0	0	30
P3.4	part./100 g	0	—	20	—	0	—	20	—
	part./kg	0	—	200	—	0	—	200	—
P3.5	part./100 g	0	—	4	—	0	—	4	—
	part./kg	0	—	40	—	0	—	40	—
P4.1	part./100 g	0	0	0	0	0	0	0	0
	part./kg	0	0	0	0	0	0	0	0
P4.2	part./100 g	0	0	0	0	0	0	0	0
	part./kg	0	0	0	0	0	0	0	0
P4.3	part./100 g	0	0	1	0	0	0	1	0
	part./kg	0	0	10	0	0	0	10	0
P5.1	part./100 g	30	13	6	10	0	0	36	23
	part./kg	300	130	60	100	0	0	360	230

		Fragments		Fibres		Other		Total	
		2018	2019	2018	2019	2018	2019	2018	2019
P5.2	part./100 g	13	5	32	13	0	0	45	18
	part./kg	130	50	320	130	0	0	450	180
P5.3	part./100 g	8	17	12	0	1	0	22	17
	part./kg	80	170	120	0	10	0	220	170
P6.1	part./100 g	86	17	430	26	0	0	516	43
	part./kg	860	170	4300	260	0	0	5160	430
P6.2	part./100 g	9	3	82	11	0	0	91	14
	part./kg	90	30	820	110	0	0	910	140
P6.3	part./100 g	5	—	31	—	0	—	36	—
	part./kg	50	—	310	—	0	—	360	—
P6.4	part./100 g	12	—	42	—	0	—	55	—
	part./kg	120	—	420	—	0	—	550	—

Source: Author (2020).

Table 1.
Average number of artificial micro-materials observed in the sediment in the study area (sorting with a stereomicroscope, referring to grain sizes between 75 μm and 1.18 mm). Where: -, non-existent layer and 0, no fragments.

The results indicate that the Água Branca stream basin is undergoing a process of artificialisation of the sediment of urban water bodies, with microplastics being the most significant materials incorporated into the sediment profile.

The sedimentation and accumulation of materials visible to the naked eye (macro fragments) such as cardboard, plastics and the remains of construction materials can be seen in the stream beds.

The Tibiriçá Reservoir - P6 (a lentic environment with a slower water flow) has the greatest accumulation of micro and macroplastics, as shown in **Figure 3**. This site receives all the drainage from the urban area of Itirapina, which is part of the Água Branca Stream basin, and seems to contain the macroplastics. It was possible to collect packaging dating back to 1993, showing the degradation caused by poor solid waste management.

Of the material observed, the lowest concentration is in the deepest layer of the sediment, indicating greater removal of light materials by the water or greater recent input of anthropogenic materials.

In the study, we observed that a higher percentage of artificial materials occurs in the November 2018 collection compared to May 2019. It may be related to the intense rainy season prior to the last collection, which may have carried a greater contribution of mineral sediments and carried away the fragments of microplastics, which are the main materials found.

We concluded that the type and intensity of land use influences the amount of microplastics. Considering the watercourses (lotic environments) in the tributary of the Água Branca Stream in an urban area with a lower density of occupation and surrounding grass vegetation (P1), we found a lower amount than at the point in the urban centre (P2), surrounded by commercial and residential use and without riparian vegetation. We observed the highest concentration of artificial materials at this point. The point located in a rural area had less artificial material than the other two.

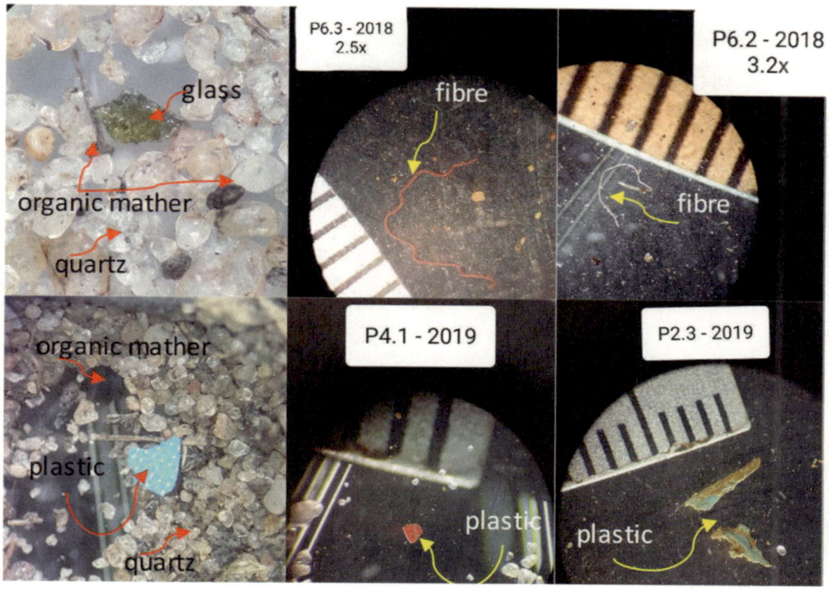

Figure 3.
Fragments of artificial materials observed in the 2018 samples in the sediment of the bodies of water in the Água Branca Stream basin. Scale: 1 mm. Source: Author (2020).

This highlights the need for intervention to prevent this artificialisation from intensifying as the city grows. Aquatic invertebrates and fish consume microplastics, and in turn, they serve as food for humans (knowledge about the deleterious effects on humans is still limited). An action to clean the reservoir and remove large plastics would be important to prevent them from degrading and generating new microplastics in the future.

4. Barrinha stream basin

The aim of this study was to identify the environmental quality and geoenvironmental changes of a watershed between a rural area and an area of environmental protection, by means of a set of environmental data and verification of the influence of land use on watercourses through deposited sediments.

The Barrinha Stream basin, with an estimated area of 7.7 km^2, in the district of Cachoeira de Emas, municipality of Pirassununga located in the interior of São Paulo state, Brazil, is a micro-watershed that is under the influence of surrounding uses (**Figure 4**).

Although much of it is within the area of the National Centre for Research and Conservation of Continental Aquatic Biodiversity - CEPTA - ICMBio, there are some springs occupied by various agricultural and urban uses, which are impacting soil degradation, sediment input into the stream, natural vegetation and, consequently, local aquatic life. Although it is not a protected area, the farm area has fragments of cerrado (savannah), cerradão (forested savannah) and riparian vegetation that are well preserved and where environmental education activities occurs.

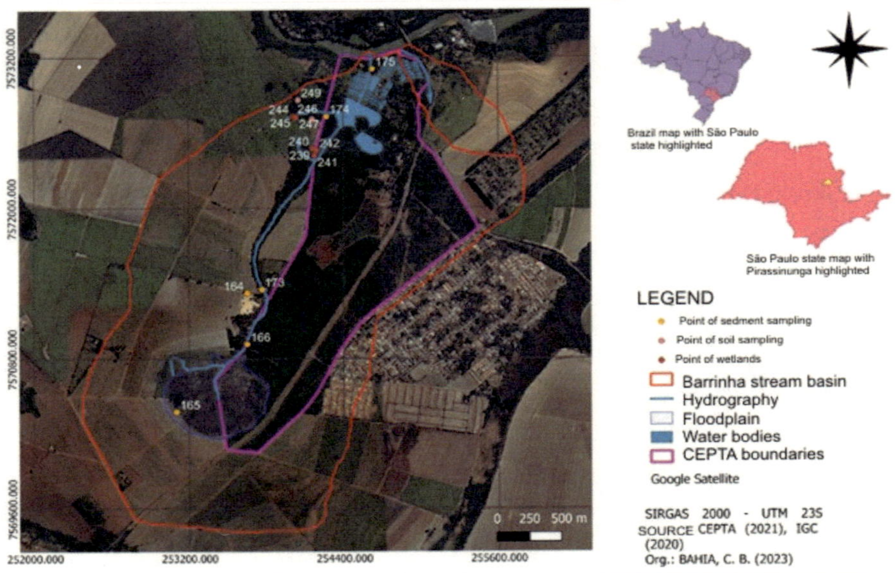

Figure 4.
Micro-basin of the Barrinha stream with the delimitation of the CEPTA area, study points and sample collection. Sources: [30, 31].

The Barrinha Stream is a second order watercourse and was originally 3180 m long. However, after the installation of five dams, embankments and the construction of tanks, around 27% (approximately 877 m) of the original length was affected [31]. The institution's fishing ponds uses the water.

The collection methodology used a corer sampler to verify the composition and characteristics of the sediments, identifying possible points of sediment accumulation and siltation related to the degradation of the basin's soil and its variability in depth, which would be related to the periods of greater and lesser water flow in the channels.

The sediments of the Barrinha stream were collected in two periods, one in the autumn of 2022 (May) after the rainy season and the second in the middle of the rainy season in the summer of 2023 (March).

The samples were collected at 4 points (P166, P173, P174, P175), which run from the floodplain of the spring to near the mouth of the stream (**Figure 5**).

As a result of this study, we observed great variability in water flows, with some stretches already characterised as intermittent.

The area of the course's main source, in a floodplain, is dry almost all year round. The influence of the surrounding sugar cane plantations is evident. It was not possible to collect recent sediment at this site.

In May 2022, the Barrinha Stream had little water flow and some of the points studied were dry or with little water. In March 2023, at the end of a period of intense rainfall, the stream had a greater quantity of water, but in some places, such as P174, there was already a significant accumulation of organic matter, indicating that the transport of mineral materials was not happening with such intensity and there was already a decrease in water flow, although the tributary that flows into this point had a good water flow. P175, at the mouth of the Barrinha Stream on the Mogi Guaçu River,

Point, coordinates (WGS 84) and description		
	May 2022	**March 2023**
P166 – 253649; 7570917 Close to a rural property.	Small stretches of bed with water, some excavated 	Water throughout the stretch, with little flow and a lot of OM 
P173 – 253763;7571355 Downstream from the mining area, on a rural property. There are two dams in the area (both with dry reservoirs)	Dry stream, but the soil was damp. 	Stream with a large amount of water. Construction debris on the road nearby 
P174 – 254270;7572729 Point of confluence with a tributary	Little water flow, constant in the Barrinha and in the tributary 	Constant flow, large quantity of OM in the Barrinha and more sediment in the tributary 
P175 – 254626;7573106 Near the mouth of the Córrego da Barrinha on the Mogi Guaçu River, at the point of confluence with the water coming out of the CEPTA tanks	Constant flow, shallow water depth 	Water depth > 1 m due to flooding in the Mogi Guaçu River 

Figure 5.
Sediment collection points, UTM WGS 84 coordinates (m). Description of the points in May 2022 and March 2023. Source: Authors, 2022.

had a rise in level of approximately 1 m, influenced by the flooding of the latter, and it was not possible to collect profile 2.

The samples collected measured between 12 and 31 cm in length. We separated the profiles collected (**Figure 6**) into 1 to 4 sub-samples samples by homogeneous sections (colour, texture, fraction), that we sent to the laboratory for analysis.

The fraction separation and treatment procedures were the same as those described in item 2, with some modifications because there were points with a large accumulation of macroscopic organic matter (branches and leaves, for example) and also because the sediment was finer and more aggregated.

The samples were left to dry at room temperature and, for the silt-clay samples that were more difficult to dry, we used an open oven. The use of a closed oven is not recommended because there is a risk of burning other materials present in the sample, such as anthropogenic materials.

After drying, the samples were first analysed visually, looking at the content of the organic matter fragments and their size. Then we weighed the samples and the visible organic matter (e.g. twigs, seeds, leaves, small insects) separated.

In this study, there were many samples with a large amount of OM and we used degradation. The method promotes the removal of organic colloids and the dispersion of clay, facilitating the granulometric separation of the fractions [32].

We analysed 16 samples from Barrinha Stream in May 2022 and 12 samples in March 2023 (**Table 2**).

Profile 1 of P175, which had been subdivided into three samples in the first collection, showed only one sample of homogeneous sediment in the second collection, indicating greater deposition of mineral materials, compatible with the input during the rainy season.

Analysis of the sediment samples showed that in the period without rainfall (May 2022) there was more accumulation of OM (> %), in more subsamples of the sand fraction, compatible with the period of lower water flow.

In the May 2022 collection, Am1 from P174 - profile 1 had the most organic matter, accounting for 14.7% of its total weight. Am1 from P175 - profile 1 was the sample with the lowest percentage of organic matter of its total weight, with only 1.3%.

However, in the 2023 collection, only four samples underwent the procedure. And of these four, the sample with the highest percentage of organic matter is Am 3 from P173, indicating a large input of sediment coming with the water flow and deposited on the previously dry site with a large amount of MO deposited, which corroborates the result found.

The granulometric separation (**Table 3**) showed that sandy sediments predominate (fraction between 1.18 and 0.75 mm).

Point	Image	Point	Image
P174 profile 1 Total length 12cm; Sample 1 8cm; Sample 2 4cm		P174 profile 2 Total length 18cm; Sample 1 6.5cm; Sample 2 9.5cm; Sample 3 3cm	

Figure 6.
Examples of the profiles collected in the area. Source: Photographs by the authors (2022).

	May/2022				March/2023			
Point	**Total (cm)**	**Sample**	**Length (cm)**	**% OM**	**Total (cm)**	**Sample**	**Length (cm)**	**% OM**
p. 166	31	1	17		25.5	1	12	4.82
		2	5			2	10.5	5.64
		3	5			3	3	7.79
		4	4	2.5				
p. 173	No profile taken (=40)	1		9.17	23	1	3	
		2		5.3		2	10	
		3		2.33		3	7	
						4	3	8.84
p. 174 (perfil 1)	12	1	8	14.7	22	1	6	
		2	4	3.1		2	16	
p. 174 (perfil 2)	18	1	6.5		22	1	17.5	
		2	9.5			2	4.5	
		3	3					
p. 175 (perfil 1)	21.5	1	6.5	1.3	15	1	15	
		2	13					
p. 175 (perfil 2)	15	1	7.5		No collection due to excess water			
		2	7.5					

Where: sample 1 = top; samples 2, 3 and 4 = > depth.

Table 2.
Measurement of the samples collected in May 2022 and March 2023 and % MO of the samples that underwent degradation with H_2O2.

The samples from P175, profile 1/Am2/2022 and profile 2/Am2/2022 had fragments of basaltic and sandstone rocks and a greater weight of the >4.75 mm fraction.

Stereomicroscope analyses revealed that there were not many fragments of anthropogenic materials.

The May/2022 samples had the same basic composition: quartz with some iron oxide concretions. Samples P165 - Am1 (floodplain area) and P166 - Am4 had a large amount of OM (roots, branches, seeds and bark) in all fractions. Sample P166 - Am3 had a fibre that appeared to be fabric and sample P175 (profile 1) - Am 2 had a fragment of foam.

In the March/2023 samples, the composition remained the same. However, we observed more rounded grains of quartz, indicating a possible different source of the material. Samples P173 - Am2, P174 (profile 1) - Am1 and P174 (profile 1) - Am2 contained a large amount of OM, such as seeds or twigs, in all their fractions. Sample P174 (profile 1) - Am1 had a fragment that looked like cardboard. Sample P174 (profile 2) - Am1 had a plastic fragment (**Figure** 7).

The presence of quartz grains in all the samples reiterates the perception that the micro-basin has mainly sandy-textured soil.

The use of the surrounding land (crops, mining and dams) has negatively affected the environmental quality of the watershed. Temporary crops such as sugar cane mostly occupy the basin. The areas with anthropogenic influence reduce the flow of surface water, as in the reduced floodplain due to nearby temporary crops, and

Point	Sample (Am)	Total sample weight (g)	Fraction weight > 4.75 mm (g)	Fraction weight between 4.75 and 1.18 mm (g)	Fraction weight between 1.18 and 0.75 mm (g)	Fraction weight < 0.75 mm (g)
May 2022						
166	1	29.93	0	1.44	27.81	0.53
	2	30.35	0	3.85	25.87	0.61
	3	31.04	0	0.61	29.57	0.65
	4	29.14	0	0.27	27.83	0.54
173	1	21.81	0	0.16	15.80	5.85
	2	26.81	0	<0.001	14.55	11.91
	3	28.99	0	<0.001	25.46	3.46
174 (prof 1)	1	6.94	0	<0.001	4.44	2.42
	2	35.74	0	0.45	32.02	3.48
174 (prof 2)	1	30.40	0	<0.001	29.70	0.35
	2	30.57	0	0.66	29.54	0.28
	3	30.84	0	0.87	29.25	0.33
175 (prof 1)	1	31.36	0.69	0.45	32.02	3.48
	2	171	96.04	2.44	64.98	7.37
175 (prof 2)	1	37.01	8.13	1.50	25.30	2.23
	2	196	104.83	15.29	63.49	3.09
March 2023						
166	1	29.21	0	0.009	27.71	1.35
	2	28.93	0.87	2.23	23.57	2.09
	3	28.29	1.6	2.93	20.40	2.52
173	1	18.55	4.78	7.4	5.5	0.89
	2	30.48	0	2.2	25.31	3.2
	3	27.75	0	0.009	23.3	4.1
	4	30.28	0	0.14	26.68	2.93
174 (prof 1)	1	31.29	3.35	8.36	17.57	0.66
	2	30.18	0.26	2.12	25.5	1.8
174 (prof 2)	1	30.59	0	0.62	29.76	0.001
	2	30.51	0	1.28	28.28	0.85
175 (prof 1)	1	30.39	0	0.89	24.82	4.05

Table 3.
Weight of the samples collected and their respective fractions.

generate an increase in sediment in the watercourses. The preservation of forest vegetation is greater within CEPTA's boundaries, favouring the availability of water and the balance of the streams, such as the spring that flows into P174. We also observed eroded areas and exposed soil used for dumping waste, which are source areas for the sediment that reaches the stream.

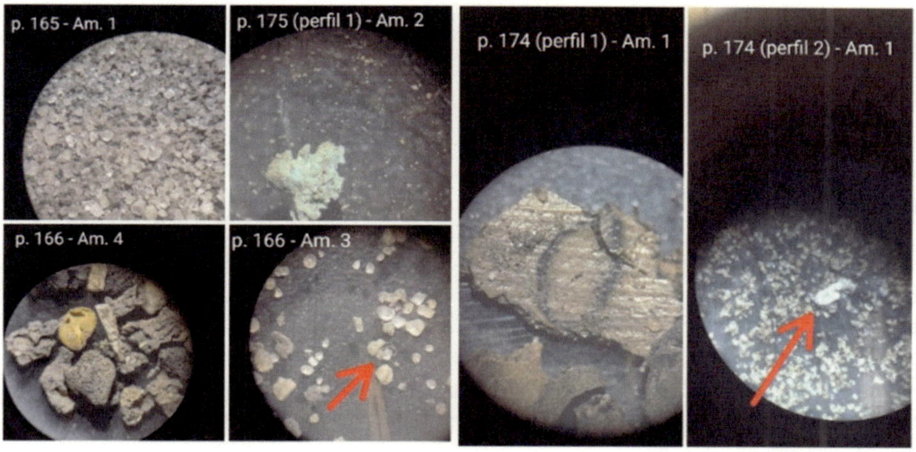

Figure 7.
Photographs of items highlighted in the microscopic analysis. Source: Author (2023).

In conclusion, it is clear that the use of the area around the Barrinha Stream has significantly affected its watercourse. We found many points with sediment accumulation, raising the concern of possible silting of the water body, combined with the low water flow, which prevents sediment transport. We found traces of materials of human origin in sediment samples, as well as the greater degradation of water bodies where there is construction in the surroundings. The floodplain area is almost permanently devoid of water accumulation, and organic soil has shown that there has been a reduction in its perimeter, influencing the flow of the Barrinha Stream, which is intermittent at various points where there was a continuous flow observed in previous research in 1994 [33].

The area of forest on CEPTA's perimeter and its effect on the perenniality of the watercourses highlight its importance for preserving native vegetation in the basin, in contrast to the surrounding areas where cultivated land predominates.

It is necessary to invest in measures to protect and conserve the watercourses in order to minimise the negative impacts of the use of the surroundings, especially the springs.

5. Discussion

The formation of river sediments with anthropogenic alteration occurs when artificial materials are incorporated into the riverbed. Plastics are the main solid waste generated in urban centres, so they are the main waste incorporated into sediments.

There is a great deal of media coverage of the harmful effects of plastics on the aquatic environment, degrading not only the quality of this resource but also the biota. Despite being a major current environmental problem, research into the impacts of this material is also recent, especially if we consider microplastics (particles smaller than 5 mm) [34].

According to the survey carried out by [35], benthic fauna can consume microplastic incorporated into river sediment depending on the organisms' feeding mode. In Serra Talhada, Pernambuco, Brazil, [36] observed plastic residues in the stomach contents of 83 per cent of the fish collected. These researchers state that their study suggests that the biota is vulnerable to the microplastics present in freshwater and that they are associated with areas with more intense urbanisation, also highlighting that the poorest community feeds on these fish.

With regard to the harmful effects of microplastics in drinking water, the World Health Organisation (WHO) [37] states that the knowledge available to date is limited and represents a low concern for human health. On the other hand, this same lack of conclusive studies is a cause for concern, because it is still unknown how these particles act on our bodies.

The WHO points out that plastics have low toxicity and, because they are insoluble, are unlikely to be absorbed by the gastrointestinal tract. However, it also points out that particle size can influence the absorption of toxicity, with smaller particles posing a greater risk. He also points out that additives and monomers (molecules made up of a single segment) can become bioavailable in the gastrointestinal tract [37]. The human body exposes itself to microplastics through ingesting food or water containing microplastics, inhaling microplastics in the air or dermal contact through textiles or air [38].

As pointed by [39], each person ingests 39,000 to 52,000 microplastic particles per year, depending on the age and gender of the individual. If we consider inhalation, this value varies between 74,000 and 121,000 particles per person. Also according to these authors, if we consider the recommended water intake using only bottled water, a consumption of 90,000 microplastic particles occurs, while if only tap water is consumed there are 4000 particles (research carried out considering a North American diet).

According to [38], the increase in the occurrence of neurodegenerative diseases, immunological disorders and cancer may be related to increased exposure to environmental contaminants, including microplastics. Therefore, knowing the distribution pattern of these materials and quantifying them is of the utmost importance to subsidise future studies related to research into the harmful effects on humans.

From a geological point of view, studies suggest that the artificialisation of sediment can change the patterns of sediment erosion, transport and deposition. An initial experiment carried out in the laboratory considered different microplastics (types, shapes and sizes) compared to natural sediment. Microplastics with spherical shapes move more quickly compared to natural sediments [40].

The results obtained in the studies are in line with the global trend of increasing concentrations of artificial materials incorporated into river sediments.

The bibliographical research showed the lack of studies carried out in tropical freshwater river environments, especially studies of layer profiles. In this sense, our research helps to understand the behaviour of riverbed sediment anthropisation.

Until now, there are no standardised methods for detecting microplastics in sediments, which makes comparisons difficult.

Table 4 summarises the main research carried out on the subject, relating to sediment collection methodologies and analysis of the results.

With regard to the sampling methodology and its results, we observed that studies are converging towards obtaining in-depth data, with profile sampling, emphasising the relevance of the methodology proposed in this chapter. The results we obtained are consistent with those obtained in other studies [21, 22, 49, 50].

The degradation of the organic matter in the samples proved to be important in facilitating the visualisation and optical identification of the artificial materials incorporated into the sediment in the two cases studied. In the second case, in an area

Author	Year	Goals	Results
[41]	2017	Whether a WWTP effluent is a source of MPs in a river in Canada and whether this material is deposited in the sediment, with an Ekman sampler	They found few particles in relation to other studies and emphasise the need for similar methods for comparison
[42]	2019	Study the types and sizes of plastic fragments in the Guanabara basin (Rio de Janeiro, Brazil) using a Van Veen grab	The high availability of MPs in the sediment suggests high risk of contamination to benthic organisms and fishes
[43]	2019	Study the incidence of MPs as a function of population size and urbanisation in Auckland (New Zealand)	In the MP incidence, local-scale factors may be more important than catchment-scale processes in small streams
[44]	2019	Investigate the abundance, distribution and sources of MPs in Tibetan river sediments collected with stainless steel shovel	The abundance of MP in the sediments of urban areas was higher than that in rural areas. But there was no statistical difference in the surface water of urban and rural areas
[45]	2019	Estimate MP values in the water column and surface sediments in Florida's largest estuary (Tampa Bay), including spatial and temporal variations in these values	Found highest concentrations of MP in sediments close to industrial sources. It is impractical to remove the numerous and tiny microplastic particles from the water or sediments
[27]	2019	Investigate the abundance of MPs in a highly urbanised area in Mexico, using a trowel and a Van Veen grab sampler	MPs were more abundant in the urban-industrial corridor
[46]	2019	Study the sources and potential risks of MPs in Lake Poyang (China)	The results indicate domestic sewage and fishing activities as the main sources of MP found in water, sediments and biota
[47]	2020	Check for a correlation between local characteristics and the abundance of MPs in sediments of a river in Canada, using a Petite Ponar Grab for sampling	Concentrations of MPs are due to environmental filters, point sources and sediment characteristics
[48]	2018	Establish a methodology to assess the risk of microplastic contamination in sediment for the context of urban rivers of Chinese mega cities	Using the plastic compounds, they defined the toxicity and, after the recognition of the MP type and concentration, defined the risk of contamination
[4]	2018	Establish the magnitude of contamination of benthic sediments by MPs and explain the observed patterns in abundance and distribution in rivers in the United Kingdom	The biggest change in MP abundance was due to the presence of a lake, more than to population density. They observed a mechanism of MP fixation in freshwater environments by caddisflies
[22]	2019	Investigated sources, fate and seasonal and spatial distribution of MPs in a river basin and estuary in China, investigating water and sediment. Used a grasp bucket for sediment	Areas with high population and gross domestic product presented more MPs in sediment and water, indicating impacts of human activities and economic activities
[21]	2017	Study on the pollution of sediments in water bodies due to land use	More than the type of use, the complexity of the forms of occupation and layout of the different land uses influence the distribution of pollutants
[49]	2021	Study of MPs in sediment layers using profile sampling in a river basin in China	The plastic particles decrease in depth and occur more frequently in places with more intense human activity

Author	Year	Goals	Results
[50]	2024	Reviewed studies about MPs in Asian lentic environments	The lakes are overlooked and poor studied. It is not possible to compare results due to absence of standard methods
[10]	2024	They studied the sediments upstream and downstream of the French metropolis, with collection of sediment profiles and dating	There is a greater volume of MP downstream and also in more recent layers, but with a decrease in polymers from industrial sources

Table 4.
Summary of the main objectives and results of the most recent research into microplastics (MP) in sediments.

with a higher content of riparian vegetation, it was crucial to separate the humidified OM from the mineral grains and other fragments.

As [45] states, it is impractical, if not impossible, to remove microplastics from the water column or separate this contamination from sediments and that the best way to reduce the potential risks of this contamination is to control the source of microplastics entering the aquatic system. The environmental legislation aimed at restricting plastics may be responsible for the reduction in some types of plastic found in some studies [10].

6. Conclusion

There are various methods of sediment collection used to study anthropisation and detection of non-natural particles, such as microplastics, and it is not possible to make a complete comparison of the results found or even of the most correct methodology [23, 50].

In this research, we developed the geomechanical PVC sampler, which fulfilled its functions and proved to be a viable and low-cost tool, but it is necessary to discard from the fragment counts all the possible particles of this material found in the sediments. Other studies have used samplers with transparent PVC liners, with the same problem of the possibility of contamination [10], but the analysis of the volume collected per layer is lost when the aim is to evaluate particles per volume collected. With the sampler used, it was possible to take sediment samples of a significant size (from 16 to 26 cm in the first study and from 12 to 31 cm in the second), within the range of values obtained in other studies [51].

The study of sediment can lead to an understanding of the degradation caused to the aquatic system and the sediment profile can show us the impacts suffered by the water body over time [52]. In the two cases studied, we observed this degradation.

In the Água Branca Stream basin, the urban influence left this record in the microplastic content, and the influence of lentic areas on retention was clear from the macroplastics found and their dating. Although solar radiation and water play the role of degradation, considering the intact materials deposited there since the 1990s, contamination by microplastics will be long-lasting. Considering the maximum values of 5160 part./kg in the 2018 sampling in P6 and comparing it with profile data from different areas from the mainland to the mouth, analysed by [51], these are compatible with some lakes and floodplains, although there are others with much higher levels and this is related to the occupation characteristics.

With regard to the rural area studied, we saw that in the sediments of Barrinha Stream, there are few plastic fragments, but the reduction in water and the large amount of sediment contributed by the replacement of natural vegetation with temporary crops are evident. For this site, studies of physico-chemical parameters may reveal other influences of agricultural activities on sediment quality. It was possible to observe that, even when the volume of anthropogenic fragments is not significant, the sediment analysis records the variations in flow and sediment input that occur over time due to occupation of the surroundings.

The knowledge and impacts of waste in the ocean is widespread, but there is still no good knowledge of these impacts in freshwater environments; likewise, research in this field is also recent, and there are no conclusive studies on the impact of microplastics on the human body, for example.

In the last years, several articles have been published dealing with continental sediments, but there is still no unified methodology and since this is a new field of study, all new approaches are necessary and important. The generation of standardised methodologies and norms happens after many different studies and methods, arriving at a robust state of the art.

The methodological procedures for sediment collection and analysis presented in this chapter and used in the two case studies possibilited in-depth analysis of the content of anthropogenic particles in the sediments, highlighting urban waste and activities as the main source of these materials.

Future studies of sediments must deal with their dating, as carried out by [10] for a more precise assessment of the inputs and removals of microplastics by water during hydrological cycles.

There is a need to expand and disseminate legislation that defines more restrictions on the use of plastics, especially those aimed at controlling disposable utensils.

It is therefore important to emphasise the importance of Environmental Education and scientific dissemination projects on the subject and to bring knowledge to the community, sensitising them to the degradation of aquatic environments and the ingestion of microplastics by invertebrates and fish and the possible impacts on human health. It is also important to emphasise the damage caused to watercourses by changing land uses; the human population grows and loses water sources necessary for its survival.

Acknowledgements

This study was financed by: the Coordenação de Aperfeiçoamento de Pessoal de Nível Superior - Brasil (CAPES) - Finance Code 001; grant The National Council for Scientific and Technological Development (CNPq) PIBIC – UFSCar; grant 2022/01276-9, São Paulo Research Foundation (FAPESP) FAPESP.

Conflict of interest

The authors declare no conflict of interest.

Author details

Denise Balestrero Menezes[1*], Daniele Toyama[2] and Clara Bianchi Bahia[3]

1 Federal University of São Carlos (UFSCar), Department of Civil Engineering, Program in Urban Engineering (PPGEU), São Carlos, Brazil

2 Urban Engineering, Program in Urban Engineering (PPGEU / UFSCar), São Carlos, Brazil

3 Federal University of São Carlos (UFSCar), São Carlos, Brazil

*Address all correspondence to: denisebm@ufscar.br

References

[1] Jesus F, Tremblay LA. Key challenges to the effective Management of Pollutants in water and sediment. Toxics. 2022;**10**:219. DOI: 10.3390/toxics10050219

[2] Aswathanarayana U. Geoenvironment: An Introduction. Rotterdam: A.A. Balkema/Brookfield; 1995

[3] Fendrich R et al. Drenagem e controle da erosão urbana. 3rd ed. São Paulo: Ibrasa. Curitiba: Champagnat; 1991. 442 p

[4] Tibbetts J et al. Abundance, distribution, and drivers of microplastic contamination in Urban River environments. Water. 2018;**10**(1):10-14. DOI: 10.3390/w10111597

[5] Zalasiewicz J, Gabbott S, Waters C. Plastic waste: How plastics have become part of the Earth's geological cycle. In: Letcher TM, Vallero DA, editors. Waste. 2nd ed. Amsterdam: Academic Press, Elsevier Inc.; 2019. pp. 443-452. DOI: 10.1016/B978-0-12-815060-3.00023-2

[6] Müller G, Middelburg JJ, Sluijs A. Introducing GloRiSe – A global database on river sediment composition. Earth System Science Data. 2021;**13**:3565-3575. DOI: 10.5194/essd-13-3565-2021

[7] Rotta CMS, Zuquette LV. Assessment of environmental degradation due to anthropogenic processes based on critical zones: A study in a basin in southern Brazil. Environmental Earth Sciences. 2021;**80**:215. DOI: 10.1007/s12665-021-09475-z

[8] Peluso ML, Demetrio PM, Bernasconi C, Aprigliano N, Riviello López G, Abelando M, et al. ¿Existe una relación entre la calidad de los cursos de aguas superficiales asociadas a los paisajes rurales y el nivel y el tipo de antropización en la zona de influencia? Ecología Austral. 2023;**33**(3):821-838. DOI: 10.25260/EA.23.33.3.0.2023

[9] Martínez-Sifuentes AR et al. Pérdida de suelo y modificación de escurrimientos causados por el cambio de uso de la tierra en la cuenca del río Conchos, Chihuahua. Nova Scientia. 2020;**25**(12):00005. DOI: 10.21640/ns.v12i25.2321

[10] Dhivert E, Pruvost J, Winiarski T, Gasperi J, Delor-Jestin F, Tassin B, et al. Time-varying microplastic contributions of a large urban and industrial area to river sediments. Environmental Pollution. 2024;**347**:123702. DOI: 10.1016/j.envpol.2024.123702

[11] Brito RNR, Asp NE, Beasley CR, Santos HSS. Características Sedimentares Fluviais Associadas ao Grau de Preservação da Mata Ciliar - rio Urumajó, Nordeste Paraense. Acta Amazônica. 2009;**39**(1):173-180. DOI: 10.1590/S0044-59672009000100017

[12] Williams M, Zalasiewicz J, Desorgher M. Underground Rivers of the Anthropocene. In: Bureau C, editor. Zürich: Lars Müller Publishers; 2023. pp. 62-71. Available from: https://www.lars-mueller-publishers.com/cave-bureau [Accessed: April 20, 2024]

[13] Zalasiewicz J. Disputed start dates for Anthropocene. Nature. 2015;**7548**(520):436. DOI: 10.1038/520436b

[14] Anthropocene Working Group – AWG. What is the Anthropocene? – Current Definition and Status. [Internet]. 2019. Available from: http://www.quaternary.stratigraphy.org/working-groups/anthropocene/ [Accessed: 04 January 2024]

[15] Waters CN et al. The Anthropocene is functionally and stratigraphically distinct from the Holocene. Science. 2016;**351**(6269):137-147. DOI: 10.1126/science.aad2622

[16] Associação Brasileira da Indústria do Plástico - ABIPLAST. Indústria Brasileira de Transformação e Reciclagem de Material Plástico - Perfil 2017. 2017. 87p. Available from: https://www.abiplast.org.br/wp-content/uploads/2019/03/Perfil-2017.pdf [Accessed: 06 January 2024]

[17] FUNVERDE – Organização Não Governamental. A evolução do plástico – A origem das sacolas plásticas. [Internet] s/d. Available from: https://www.funverde.org.br/blog/sacolas/a-evolucao-do-plastico/ [Accessed: 07 January 2024]

[18] Viana MB. Sacolas plásticas: Aspectos controversos de seu uso e iniciativas legislativas. Study. Brasília: Câmara dos Deputados – Consultoria Legislativa; 2010. 24p

[19] Zalasiewicz J et al. The geological cycle of plastics and their use as a stratigraphic indicator of the Antropocene. Anthropocene. 2016;**13**: 4-17. DOI: 10.1016/j.ancene.2016.01.002

[20] Moore CJ. Synthetic polymers in the marine environment: A rapidly increasing, long-term threat. Environmental Research. 2008;**108**:131-139. DOI: 10.1016/j.envres.2008.07.025

[21] Liu A, Duodu GO, Goonetilleke A, Ayoko GA. Influence of land use configurations on river sediment pollution. Environmental Pollution. 2017;**229**:639-646. DOI: 10.1016/j.envpol.2017.06.076

[22] Fan Y, Zheng K, Zhu Z, Chen G, Peng X. Distribution, sedimentary record, and persistence of microplastics in the Pearl River catchment, China. Environmental Pollution. 2019;**251**:862-870. DOI: 10.1016/j.envpol.2019.05.056

[23] Uddin S, Fowler SW, Uddin MF, Behbehani M, Naji A. A review of microplastic distribution in sediment profiles. Marine Pollution Bulletin. 2021;**163**:111973. DOI: 10.1016/j.marpolbul.2021.111973

[24] Toyama D, Fernandes V, Christoforo AL, Menezes DB. The artificialization in the sediment profiles of the streams in the Água Branca basin - Itirapina, São Paulo, Brazil. Journal of Environmental Management. 2021;**290**:112610. DOI: 10.1016/j.jenvman.2021.11261

[25] Bahia CB. O meio físico e alterações geoambientais da bacia hidrográfica do Córrego da Barrinha na área do CEPTA, Pirassununga - SP. [Scientific Iniciation]. São Carlos: Universidade Federal de São Carlos; 2023

[26] Toyama D, Menezes DB. Macro e microplástico no sedimento fluvial na cidade de Itirapina/SP, II Simpósio Nacional de Gestão e Engenharia Urbana. Blucher Engineering Proceedings. 2019;**6**:682-690. DOI: 10.1016/singeurb2019-92

[27] Shruti VC, Jonathan MP, Rodriguez-Espinosa PF, Rodríguez-Gonzalez F. Microplastics in freshwater sediments of Atoyac River basin, Puebla City, Mexico. Science of the Total Environment. 2019;**654**:154-163. DOI: 10.1016/j.scitotenv.2018.11.054

[28] Empresa Brasileira de Pesquisa Agropecuária - EMBRAPA. Manual de Métodos de Análise de Solo. 2nd ed. Rio de Janeiro: Embrapa; 1997. 212 p

[29] The Brazilian Institute of Geography and Statistics – IBGE. Cities: Itirapina. Available from: https://www.cidades.ibge.gov.br/brasil/sp/itirapina/panorama [Accessed: 07 January 2024]

[30] IGC - Instituto Geográfico e Cartográfico. Hidrografia Região Leste do Estado de São Paulo 1:25.000 (Preliminar). 2010. Available from: https://www.datageo.ambiente.sp.gov.br/app/?ctx=IGC#INSTIT [Accessed: 02 April 2020]

[31] CEPTA. A bacia de drenagem, hidrografia e outros temas correlatos do córrego da Barrinha, afluente do rio Mogi-Guaçu, Pirassununga/SP, local sede do CEPTA. In: Tec. Inf. N.4; 2021 - CEPTA/DIBIO/ICMBio. 2021

[32] Verdade FC. Ação da água oxigenada sobre a matéria orgânica do solo. Bragantia. 1954;**13**(24):287-295

[33] Nascimento VMC. Caracterização limnológica e efeitos dos vários usos do solo na qualidade da água do Córrego da Barrinha, Pirassununga – SP. [dissertation]. Brasilia: Universidade de Brasília; 1994

[34] Koelmans AA. Microplastic as a vector for Chemicalns in the aquatic environment: Critical review and model-supported reinterpretation of empirical studies. Environmental Science & Technology. 2016;**50**(7): 3315-3326. DOI: 10.1021/acs.est.5b06069

[35] Redondo-Hasselerharm PE et al. Microplastic effect thresholds for freshwater benthic macroinvertebrates. Environmental Science & Technology. 2018;**52**(4):2278-2286. DOI: 10.1021/acs.est.7b05367

[36] Silva-Cavalcanti et al. Microplastics ingestion by a common tropical freshwater fishing resource. Environmental Pollution. 2017;**221**:218-226. DOI: 10.1016/j.envpol.2016.11.068

[37] World Health Organization - WHO. Microplastics in Drinking-Water. Technical Document. 2019. 129p. Available from: https://www.who.int/publications/i/item/9789241516198 [Accessed: 05 January 2024]

[38] Prata JC et al. Environmental exposure to microplastics: An overview on possible human health effects. Science of the Total Environment. 2020;**702**:134455. DOI: 10.1016/j.scitotenv.2019.134455

[39] Cox et al. Human consumption of microplastics. Environmental Science & Technology. 2019;**53**:7068-7074. DOI: 10.1021/acs.est.9b01517

[40] Waldschläger K, Schüttrumpf H. Erosion behavior of different microplastic particles in comparison to natural sediments. Environmental Science & Technology. 2019;**53**(22):13219-13227. DOI: 10.1021/acs.est.9b05394

[41] Vermaire JC et al. Microplastic abundance and distribution in the open whater and sediment of the Ottawa River, Canada, and its tributaries. Facets. 2017;**2**:301-314. DOI: 10.1139/facets-2016-0070

[42] Alvez VEN, Figueiredo GM. Microplastic in the sediments of a highly eutrophic tropical estuary. Marine Pollution Bulletin. 2019;**146**:326-335. DOI: 10.1016/j.marpolbul.2019.06.042

[43] Dikareva N, Simon KS. Microplastic pollution in streams apanning in a

urbanisation gradient. Environmental Pollution, [s. l.]. 2019;**258**:292-299. DOI: 10.1016/j.envpol.2019.03.105

[44] Jiang C et al. Microplastic pollution in the river of the Tibet plateau. Environmental Pollution. 2019;**249**:91-98. DOI: 10.1016/j.envpol.2019.03.022

[45] McEachern K et al. Microplastics in Tampa Bay, Florida: Abundance and variability in estuarine waters and sediments. Marine Pollution Bulletin. 2019;**148**:97-106. DOI: 10.1016/j.marpolbul.2019.07.068

[46] Yuan E et al. Microplastic abundance, distribution and composition in water, sediments, and wild fish from Poyang Lake, China. Ecotoxicology and Environmental Safety. 2019;**170**:180-187. DOI: 10.1016/j.ecoenv.2018.11.126

[47] Crew A, Gregory-Eaves I, Ricciardi A. Distribution, abundance, and diversity of microplastics in the upper St. Lawrence River. Environmental Pollution. 2020;**260**:113994. DOI: 10.1016/j.envpol.2020.113994

[48] Peng G et al. Microplastics in freshwater river sediments in Shanghai, China: A case study of risk assessment in mega-cities. Environmental Pollution. 2018;**234**:448-456. DOI: 10.1016/j.envpol.2017.11.034

[49] Zhou Z, Zhang P, Zhang G, Wang S, Cai Y, Wang H. Vertical microplastic distribution in sediments of Fuhe River estuary to Baiyangdian wetland in northern China. Chemosphere. 2021;**280**:130800. DOI: 10.1016/j.chemosphere.2021.130800

[50] Sadia MR, Hasan M, Abu Reza M, Islam T, Jion MMMF, Masud MAA, et al. A review of microplastic threat mitigation in Asian lentic environments. Journal of Contaminant Hydrology. 2024;**260**:104284. DOI: 10.1016/j.jconhyd.2023.104284

[51] Yuan B, Gan W, Sun J, Lin B, Chen Z. Depth profiles of microplastics in sediments from inland water to coast and their influential factors. Science of the Total Environment. 2023;**903**:166151. DOI: 10.1016/j.scitotenv.2023.166151

[52] Honorato EV. Sedimentos da Lagoa Olho D'ággua: Geocronologia e acumulação de metais traços. [Thesis]. Recife: Universidade Federal de Pernambuco; 2002